Materials Chemistry of Fullerenes, Graphenes, and Carbon Nanotubes

Materials Chemistry of Fullerenes, Graphenes, and Carbon Nanotubes

Editors
Giuseppe Cirillo
Long Y Chiang

MDPI • Basel • Beijing • Wuhan • Barcelona • Belgrade • Manchester • Tokyo • Cluj • Tianjin

Editors
Giuseppe Cirillo
Pharmacy, Health and
Nutritional Sciences
University of Calabria
Rende
Italy

Long Y Chiang
Chemistry
University of Massachusetts
Lowell
USA

Editorial Office
MDPI
St. Alban-Anlage 66
4052 Basel, Switzerland

This is a reprint of articles from the Special Issue published online in the open access journal *Molecules* (ISSN 1420-3049) (available at: www.mdpi.com/journal/molecules/special_issues/Fullerenes_Graphenes).

For citation purposes, cite each article independently as indicated on the article page online and as indicated below:

LastName, A.A.; LastName, B.B.; LastName, C.C. Article Title. *Journal Name* **Year**, *Volume Number*, Page Range.

ISBN 978-3-0365-2188-6 (Hbk)
ISBN 978-3-0365-2187-9 (PDF)

© 2021 by the authors. Articles in this book are Open Access and distributed under the Creative Commons Attribution (CC BY) license, which allows users to download, copy and build upon published articles, as long as the author and publisher are properly credited, which ensures maximum dissemination and a wider impact of our publications.

The book as a whole is distributed by MDPI under the terms and conditions of the Creative Commons license CC BY-NC-ND.

Contents

About the Editors vii

Preface to "Materials Chemistry of Fullerenes, Graphenes, and Carbon Nanotubes" ix

Luca Bellucci and Valentina Tozzini
Engineering 3D Graphene-Based Materials: State of the Art and Perspectives
Reprinted from: *Molecules* **2020**, *25*, 339, doi:10.3390/molecules25020339 1

Elisa Thauer, Alexander Ottmann, Philip Schneider, Lucas Möller, Lukas Deeg, Rouven Zeus, Florian Wilhelmi, Lucas Schlestein, Christoph Neef, Rasha Ghunaim, Markus Gellesch, Christian Nowka, Maik Scholz, Marcel Haft, Sabine Wurmehl, Karolina Wenelska, Ewa Mijowska, Aakanksha Kapoor, Ashna Bajpai, Silke Hampel and Rüdiger Klingeler
Filled Carbon Nanotubes as Anode Materials for Lithium-Ion Batteries
Reprinted from: *Molecules* **2020**, *25*, 1064, doi:10.3390/molecules25051064 17

Manuela Curcio, Annafranca Farfalla, Federica Saletta, Emanuele Valli, Elvira Pantuso, Fiore Pasquale Nicoletta, Francesca Iemma, Orazio Vittorio and Giuseppe Cirillo
Functionalized Carbon Nanostructures Versus Drug Resistance: Promising Scenarios in Cancer Treatment
Reprinted from: *Molecules* **2020**, *25*, 2102, doi:10.3390/molecules25092102 37

Katerina Vrettos, Konstantinos Spyrou and Vasilios Georgakilas
Graphene Aerogel Growth on Functionalized Carbon Fibers
Reprinted from: *Molecules* **2020**, *25*, 1295, doi:10.3390/molecules25061295 69

Yi Yang, Jing Cao, Ning Wei, Donghui Meng, Lina Wang, Guohua Ren, Rongxin Yan and Ning Zhang
Thermal Conductivity of Defective Graphene Oxide: A Molecular Dynamic Study
Reprinted from: *Molecules* **2019**, *24*, 1103, doi:10.3390/molecules24061103 81

Maik Scholz, Yasuhiko Hayashi, Victoria Eckert, Vyacheslav Khavrus, Albrecht Leonhardt, Bernd Büchner, Michael Mertig and Silke Hampel
Systematic Investigations of Annealing and Functionalization of Carbon Nanotube Yarns
Reprinted from: *Molecules* **2020**, *25*, 1144, doi:10.3390/molecules25051144 91

He Yin, Min Wang, Loon-Seng Tan and Long Y. Chiang
Synthesis and Intramolecular Energy- and Electron-Transfer of 3D-Conformeric Tris(fluorenyl-[60]fullerenylfluorene) Derivatives
Reprinted from: *Molecules* **2019**, *24*, 3337, doi:10.3390/molecules24183337 105

Zeynab Zohdi, Mahdi Hashemi, Abdusalam Uheida, Mohammad Mahdi Moein and Mohamed Abdel-Rehim
Graphene Oxide Tablets for Sample Preparation of Drugs in Biological Fluids: Determination of Omeprazole in Human Saliva for Liquid Chromatography Tandem Mass Spectrometry
Reprinted from: *Molecules* **2019**, *24*, 1191, doi:10.3390/molecules24071191 123

About the Editors

Giuseppe Cirillo

Dr. Giuseppe Cirillo received his Ph.D. in Methodologies for the Development of Molecules of Pharmaceutical Interest at University of Calabria, Italy, in 2008, with subsequent 6-year postdoctoral fellows experiences between University of Calabria, Italy, and Leibniz Institute for Solid State and Materials Research Dresden, Germany. From 2010 to 2013 he was co-founder and CEO of University of Calabria spin-off company. He got the Italian National Scientific Qualification for associate professorship in Drug Technology, Socioeconomics and Regulations and Principles of Chemistry for Applied Technologies. He is currently researcher at the Department of Pharmacy, Health and Nutritional Sciences, University of Calabria, Italy. He has co-authored 118 scientific journal publication.

Long Y Chiang

Professor Long Y. Chiang received his Ph.D. degree of organic chemistry at Cornell University, Ithaca, New York, USA in 1979 with subsequent one-year postdoctoral fellow experience at Johns Hopkins University, Baltimore, Maryland, USA. He worked as a senior research staff at Corporate Research Laboratory, Exxon Research and Engineering Company, New Jersey, USA from 1981 to 1993. He assumed his Chair Research Fellow position at National Taiwan University from 1994 to 2001 and continued as a Professor at Chemistry Department, University of Massachusetts of the current position from 2001. He served as an adjunct primary professor at Department of Pathology and Laboratory Medicine, University of Toronto, Canada from 2001 to 2012. He was founded as a fellow of International Association of Advanced Materials in 2020 and a Vebleo Fellow in 2021. He has co-authored 40 technical patents and 288 scientific Journal publications.

Preface to "Materials Chemistry of Fullerenes, Graphenes, and Carbon Nanotubes"

Nanocarbon compounds, including fullerenes, graphenes, and carbon nanotubes, have attracted ever increasing interest in the materials science field owing to their superior physical, chemical, mechanical, and biological features, associated with intrinsic characteristics nonexistent in conventional organic substances. Such materials have been proposed and utilized in a large number of applications across different research and technological disciplines, extending from chemistry and engineering to subjects of molecular and polymeric electronics, photonics, materials science, energy management, and biomedicine.

This Special Issue is intended as a platform for the interactive material science articles with an emphasis on the preparation methods, functionalization chemistry, and structural characterization of nanocarbon compounds, as well as all aspects of physical properties of functionalized, conjugated, or hybrid nanocarbon materials, and their associated applications. Some recent advances in the field are collected and presented here, providing new ideas for discussion among researchers working in this multidisciplinary scenario and development.

Both recent research articles and review papers are collected, with the latter being presented at the beginning of the Book Issue since they provide an extensive overview on the topics presented herein with discussion of future perspectives.

In the first review article, Luca Bellucci and Valentina Tozzini made an overview on the strategy and preparative methods for extending two-dimensional (2D) materials of graphene in nature into the construction of 3D-architectural multilayer structures. The article outlined two main experimental approaches with synthetic routes reported for bulk preparation in reference with theoretical calculation and simulation studies of graphene-based 3D materials. These methods included the mixing for effecting intercalation process and the functionalization with pillars molecules to separate (r)GO sheet layers. It also provided the highlighting of the advantages and disadvantages of each strategy applied. Finally, they introduced the possibility of a third approach involving the use of epitaxial vapor deposition for laying regularly nano-patterned carbon buffer layer precursors prior to graphene sheet attachments, giving a suggested improvement on 3D-architectural control.

In the second review article, Thauer, et al. demonstrated the potential importance of using endohedral encapsulation of redox couplers inside multiwalled cabon nanotubes (MWCNT) to serve as hybrid carbon-based hierarchical nanostructured anode materials in the fabrication of lithium-ion batteries as electrochemical energy storage devices with high specific capacities. The approach was taking advantages from the combination of nanosize effects for increasing the surface area, enhanced electrical conductivity for facilitating the charge-transport/distribution, and better CNT-shell protection for maintaining high-performance of the device against electrode cracking and degradation of inorganic redox materials.

In the third review article, Curcio, et al. described the application of hybrid materials based on carbon nanostructures in biomedicine and in cancer treatment. Authors discussed the most relevant examples of hybrid nanosystems proposed for MDR reversal, taking into consideration the functionalization routes, as well as the biological mechanisms involved and the possible toxicity concerns.

This Special Issue also includes five research articles.

The first paper by Vrettos, et al. reported on the mechanical reinforcement of graphene aerogel with surface-functionalized carbon fibers by combining reduced graphene oxide (rGO) hydrogel and carbon fibers as highly stable 3D-porous conductive composite materials, covalently linked by epoxy and tetramines. Characterization, in terms of surface properties, as well as electrical behavior were given with suggestions of potential light-weight conductive cable uses.

The second paper by Yang, et al. described the correlation of phonon scattering induced by the degree of oxidative vacancy defects of graphene oxide to its thermal conductivity using large-scale molecular dynamic simulation approach method. It provided the theoretical guidance in possible design of tailored graphene oxide microstructures for thermal management and thermoelectric applications by varying the oxidation degree of GO or rGO.

The third paper by Scholz, et al. described high-temperature graphitized multiwalled carbon nanotube yarns made by a dry-spinning process to produce carbonaceous materials. The products may have great interest for different applications due to their superior thermal conductivity, electrical conductivity, and mechanical properties. In detail, they showed the strategies to enhance the properties of such nanostructures to be used as a replacement for common materials in the field of electrical wiring.

The fourth paper by Chiang's group described the synthesis of new 3D conformers to exhibit a nanomolecular configuration with geometrically branched 2-diphenylaminofluorene chromophores using a symmetrical 1,3,5-triaminobenzene ring as the center core. Authors stated that a nanostructure with a non-coplanar 3D configuration in design should minimize the direct contact or π-stacking of fluorene rings with each other during molecular packing to the formation of fullerosome array. The materials consisted of electron donor–acceptor conjugations intended for enhancing photoinduced intramolecular electron-transfer generation of reactive oxygen species (ROS) for photodynamic therapy applications.

The fifth paper by Zohdi and co-workers presented experimental evidence on the preparation of graphene oxide tablets based on a mixture of graphene oxide and polyethylene glycol on a polyethylene substrate to be employed for the extraction and concentration of omeprazole in human saliva samples with high efficiency.

Giuseppe Cirillo, Long Y Chiang
Editors

Review

Engineering 3D Graphene-Based Materials: State of the Art and Perspectives

Luca Bellucci and **Valentina Tozzini** *

Istituto Nanoscienze–CNR and NEST-Scuola Normale Superiore, Piazza San Silvestro 12, 56127 Pisa, Italy; luca.bellucci@nano.cnr.it
* Correspondence: valentina.tozzini@nano.cnr.it; Tel.: +39-050-509-433

Received: 24 December 2019; Accepted: 10 January 2020; Published: 14 January 2020

Abstract: Graphene is the prototype of two-dimensional (2D) materials, whose main feature is the extremely large surface-to-mass ratio. This property is interesting for a series of applications that involve interactions between particles and surfaces, such as, for instance, gas, fluid or charge storage, catalysis, and filtering. However, for most of these, a volumetric extension is needed, while preserving the large exposed surface. This proved to be rather a hard task, especially when specific structural features are also required (e.g., porosity or density given). Here we review the recent experimental realizations and theoretical/simulation studies of 3D materials based on graphene. Two main synthesis routes area available, both of which currently use (reduced) graphene oxide flakes as precursors. The first involves mixing and interlacing the flakes through various treatments (suspension, dehydration, reduction, activation, and others), leading to disordered nanoporous materials whose structure can be characterized *a posteriori*, but is difficult to control. With the aim of achieving a better control, a second path involves the functionalization of the flakes with pillars molecules, bringing a new class of materials with structure partially controlled by the size, shape, and chemical-physical properties of the pillars. We finally outline the first steps on a possible third road, which involves the construction of pillared multi-layers using epitaxial regularly nano-patterned graphene as precursor. While presenting a number of further difficulties, in principle this strategy would allow a complete control on the structural characteristics of the final 3D architecture.

Keywords: graphene-based materials; nanoporous graphene; epitaxial graphene; molecular modeling

1. Introduction

Since the experimental confirmation of its existence [1], graphene has raised great expectations because of its exceptional properties, stemming from a fortunate combination of the electronic structure of carbon, the symmetry of its lattice, and its two-dimensional (2D) nature [2]. Besides the large charge carriers' mobility and the wide-band optical response, graphene displays extremely large resistance to tensile strain associated to a very low bending rigidity [3], leading among other things to the emergence of low energy transverse phonons [4] and ripples [5]. These properties associated to the low weight have triggered the proposal of a plethora of possible applications [6–10].

With little exceptions, however, these require some sort of manipulation of the sheet: in general nano-electronics requires doping to increase the density of states at the Fermi energy or to open a gap, which can be achieved by chemical substitutions [11], introduction of adatoms [12] or defects [13,14] or structure modulation [15,16]; for photovoltaics [9] different functionalization are required, depending on the specific use proposed (anode, cathode or photoactive layer [17]). Catalysis or environmental applications, such as water filtering, generally require sheet alteration, such as perforations of tailored size [18]. Recently, a brand new branch of investigation has stemmed from graphene in-plane large mechanical strength and elasticity [3], coupled to out of plane flexibility [5]: controlled local strain would create pseudo-magnetic fields [19], besides band-gap opening and other specific electronic structure modifications [20,21] with interesting applications in nano-electronics and photonics; both in-plane (strain [22]) and out-of-plane (rippling [23]) mechanical alterations were shown to locally change chemical reactivity opening the way to controlled chemical nano-patterning [24,25]. Interestingly, all of the suggested modifications correspond to controlled disruption of the perfect symmetry of the crystal in a different way, which leads to viewing grapheme–rather than a single material–as a sort of morphable platform to build a class of slightly different materials suitable to specific purposes [26,27].

In addition to the modification of the layer, a wide range of uses needs its volumetric extension, with the requirement, however, that the 2D properties are preserved as far as possible. This is the case in applications involving storage: fuel-gas storage (e.g., H_2), or electro-chemical energy storage (supercapacitors or batteries [28]), require a large exposed surface per unit mass (or Specific Surface Area, SSA) to achieve a large gravimetric capacity (GC) [29], and the intrinsic capability of adsorbing specific substances (gas or electrolytes). Similar requirements are needed in catalysis applications [30]. Clearly, in this case the light weight of carbon and its intrinsic two-dimensionality are crucial, electric conductance is also needed in supercaps and batteries, while electronic properties can be an important added value. Finally, a number of applications related to coating can be considered as in between superficial and volumetric ones. In these, graphene-based materials must be deposited on a given surface in thin layers–but macroscopic on the atomic scale–to several purposes: protect from atmospheric agents [31], make it conductive [32] or hydrophobic [33], yet maintaining elasticity and resistance.

Indeed, preserving the needed properties and possibly enhancing or tailoring them in the 2D to 3D passage has turned out extremely complex. Up to now, two main routes were considered, both using graphene flakes as precursors. In the first, these are created by graphite exfoliation (usually after oxidation) and suspended in various solvents, resulting in a mixture of flakes with randomly distributed sizes and shapes; upon dehydration, they form 3D scaffolds with random structure and porosity [34]. These techniques, described in the next section, have the advantage of producing in cheap and scalable way a range of different 3D graphene-based nanoporous materials (GNM). The disadvantage is the high level of disorder, and the poor capability of controlling structural and mechanical properties, which are usually characterized *a posteriori*.

Building multi-layered structures separated by molecular "pillars" is considered an alternative to control the properties of the final 3D construct: theoretically, porosity and density in such structures are determined by the size and concentration of pillars, allowing the possibility of engineering the 3D structure via the pillar molecule design. Up to now, this route has been followed using organic molecules as pillars [35,36] coupled to suspended flakes, with encouraging but still not optimal results due to the difficulty of controlling the location of pillars on the randomly shaped flakes. The latest advances in this field are reviewed in Section 3.

Clearly, the optimal route should involve the control of the pillars positioning at the nano-scale on the precursor sheets, i.e., the combination of controlled chemical nano-patterning with the possibility of stacking the patterned multilayers in a controlled way. In Section 4 we illustrate the perspective to reach these objectives using the epitaxial graphene as precursor. A summary and conclusions follow in the last section.

2. Graphene-Based Nano-Porous Materials: Production and Computer Modeling

GNM are part of a broader class, the nano-porous carbons, which comprises activated carbons, carbide derived carbons, nanofoams and nanotubes, among others. While the synthesis of scaffolds with micrometer porosity has reached quite a high level of maturity thanks to the use of nano-to-micro particles as templates [37], strictly nano-porous GNMs are more difficult to produce with controlled structural characteristics. They are generally obtained with top-down techniques, using as precursors suspension of flakes. Flakes obtained from direct exfoliation of graphite (e.g., by liquid phase exfoliation [6,7]) have more regular structure and better conductive properties, and are therefore more suitable for electronics applications, but are also more expensive and difficult to obtain and handle. Therefore, for the 3D scaffold building, usually, the process starts from the oxidation of graphite to graphite oxide, e.g., by Hummer's method [38], followed by exfoliation-reduction either thermally [39], leading to Thermal Exfoliate Graphite Oxide (TEGO), or using microwaves, leading to Microwave Exfoliate Graphite Oxide (MEGO) [40,41], resulting in materials with SSA usually not exceeding 800 m^2/gr [42,43]. Samples can be subject to additional treatments, such as further reduction, or chemical activation (e.g., with KOH), which modify the edges with the result of increasing the porosity to specific pore volume (SPV) greater than 2 cm^3/g [30,44] and improving the SSA up or exceeding the graphene limit (2630 m^2/g). The result are 3D structures with randomly distributed sp^2 areas interconnected to form a tangled scaffold with pores of size ranging in the nanometer scale (Figure 1). Overall, these materials display SSA values between 500 and 3500 m^2/g, maintaining good electrical conductivity, high mechanical strength and chemical stability [45,46]. The performances as gas absorbers are basically proportional to the SSA, reaching an excess H_2 adsorption of 7% at 77 K [44]. The actual structural features, measured by SSA and PSV and some other additional parameters, such as the pore size distribution (PSD) and the mass density ρ [47] (see Table 1), depend on all the phases of the production: the exfoliation process, determining the size and shape distribution of the flakes, the reduction, influencing the intrinsic perforation and defects of the flakes, and the activation, modifying the porosity and surfaces. Consequently, the gas adsorption could in principle be tuned provided a full control of the production process is possible.

The structure control is even more crucial when GNM are proposed as storage mean in electric or electrochemical form. Being a conductor with large surface, graphene could be used as a capacitor, whose capacitance can be largely increased adsorbing electrolytes, potentially making it a super-cap [40]. To this aim, besides the already mentioned SSA directly related to capacitance and improved by activation [48], also the intrinsic capability of adsorbing electrolytes or ionic liquids becomes a key feature [49]. Therefore, though the capacitance is generally inversely proportional to porosity, the pore sizes must also be optimized based on the size of the ionic species [50,51]. A fine tuning of the porosity can also produce ion desolvation and the consequent increase of efficiency via a pseudo-capacitance effect [52]. Similar properties are required to develop materials suitable for batteries. In particular, electric conductivity and chemical stability, besides porosity are the main requisites for the electrodes for lithium-based batteries [53,54]. Finally, GNM are attractive also as gas sorters or filters for environmental applications, e.g., water or air purification and CO_2 sequestration [55].

In summary, the need for large GC for gas or electrolytes adsorption calls for large SSA and low ρ [10,55], undermining the structural stability. On the other hand, volumetric capacity (VC) increases with ρ, and the pore size must be tuned to the adsorbed fluid [26]. Clearly, the capability of finely controlling the structural parameters has a main role [56]. This task is not only difficult, but also somehow ill-defined, since the experimental structural determinations of GNM are limited to the measurement of pair distribution functions (PDF) and average pore size or at most the PSD. Computer modeling and simulations have been called into play to compensate for the lack of detailed knowledge. However, the intrinsic disorder leaves quite a large amount of under-determinacy for model building. As a consequence, models including a degree of approximation or idealization are often used. The "perforated graphene" models [44] uses flat flakes not reconstructed at the edges and/or with regularly spaced pores [57]. Other studies are even simpler, including either the ideal slit-pore geometry [58–60] or defected [61]/rippled [62] multilayers. Finally, a number of models is based on periodic 3D structures, such as the open-carbon-frameworks [63,64] or the carbon honeycomb [65].

While regular structures cannot allow a comprehensive throughput screening of the whole structural diversity landscape of GNM, clearly, the major issue in building realistic models for nanoporous scaffolds is their intrinsic disorder, difficult to include and needing large model systems to mild the effects of the boundaries or of the superimposed artificial periodicity of the model super cell. A number of computational approaches to generate disordered GNM model systems were adopted, differing in the description of the interactions between carbon atoms [66], and in the technique used to sample the structural parameters space. In the molecular dynamics (MD)-based techniques, atomistic empirical force fields (FF) are used to handle the interactions. These must be able to describe the different possible hybridization states of carbon based on the bond-order evaluated "on the fly" [67,68] and/or the formation/dissociation of the different kinds of C-C bonds [69,70]. The model generation can then proceed "bottom up", starting from a random distribution of carbon atoms in gas phase, which are subsequently subject to molecular dynamics simulated annealing cycles (heating up to 10^4 K and slow quenching [71]). Different structural morphologies can be obtained by changing the annealing conditions [72] (temperature, pressure [73] or density [74]), which is the simulation equivalent of changing the experimental conditions of production. This "from scratch" procedure is very computationally expensive, limiting the size of the model system to tens of nm, and preventing an extensive exploration of the structural parameters space and–consequently–a fine control over the resulting structures. A completely different point of view is taken in reverse Monte Carlo methods, where the atoms configurations are generated randomly and optimized until the simulated PDF matches the experimental one. In principle, the bare version of this method returns the best approximation of the inter-atomic interaction with a two-body potential as a side result, and structures compatible with it [75]. However, the nature of the C-C interaction is intrinsically many-body, therefore further restrains (geometric or energetic) are needed during the procedure [76]. This method is less expensive and can then generate larger model systems, giving good results in the meso-scale, but needs accurate structural determinations as input, which necessarily introduce an experimental bias.

To the aim of combining a modest computational effort with realism of the final model, a good strategy is using as precursors already formed graphene portions [77,78] instead of atoms. On this road, a step forward was recently done using a model building algorithm that mimics the real synthesis [55,79]. The starting point is a mixture of flakes with size and shape distributed according to the experimentally known composition of the suspension. These are mixed to reproduce the real density and allow intersections. These, the edges and the perforations are then optimized using bond order or reactive FFs, and possibly functionalized with other species, mimicking the various experimental treatments. The system is finally refined by thermalizing MD cycles. The results realistically match the PDF and can be controlled by the starting concentration/size distribution/perforation of the flakes. Using already formed flakes as precursors, not only leads to more realistic structures, but also limits the computational effort allowing the generation and extensive study of large model systems. An overview of the available disordered GNM materials, models and their characteristics is reported in Figure 1 and Table 1.

Table 1. Disordered GNM and their structural characteristics.

Precursor	Method/Treatment	SSA m^2/g	PSV cm^3/g or Avg Pore Size	Density cm^3/g	H_2 Uptake (% at 77K) or Capacitance (F/g)	Ref.
Graphite oxide	TEGO, TEGO + KOH	2300			5%	2015 [29]
Graphite oxide	TEGO + KOH	3300	2.2 (PSV)		7%	2015 [44]
Graphite oxide	TEGO + KOH	2900	1.4 (PSV)	~1	5.5%	2015 [78]
Slit pores	Modelling	5100	0.95 (PSV)	~1	6.5%	2015 [78]
Graphite	plasma-induced exfoliation	~800	~0.8 nm		2%	2016 [79]
Graphite-/diamond-like	Heating/Quenching MD simulations	600–3000	0–1.6 (PSV)	0.5–3.5		2017 [72]
activated carbon	Thermal treatment	2220	0.67 nm	1.95	5.5%	2015 [47]
Carbon atoms	Quench MD simulations	~1900	3–15 nm	~0.9	123 F/g	2019 [71]

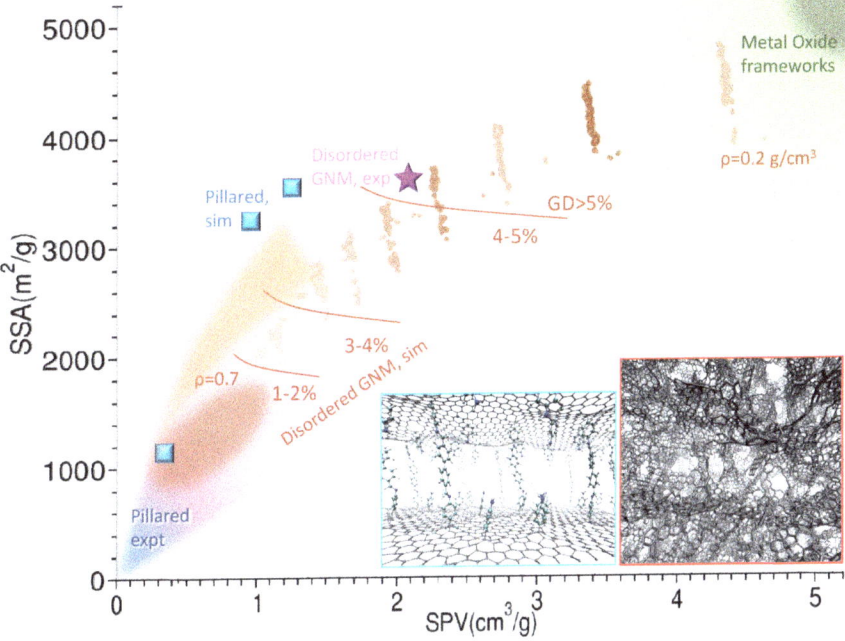

Figure 1. Structure diagram of graphene-based nanoporous materials (GNMs). The Specific Surface Area (SSA) vs specific pore volume (SPV) is reported for various experimental determinations and simulation studies. The blue shaded area encloses the experimental determinations of pillared structures, the one shaded in pink encloses the area spanned by experimental determinations of the disordered GNM scaffolds, both from various literature works cited in the text and in Table 2 (magenta star: Ref. [44]). The squared blue dots are simulations for pillared materials (Refs. [74,76,80]). Brown and reddish shades and dots are from simulations. The brown dots are preliminary from ref. [55], and roughly accumulates on lines at different decreasing density (smaller and larger simulated density are reported); red oval shade and orange shade are extracted and processed from ref. [74]. The brown lines separate areas at increasing excess GD evaluated at 77K. The region typically spanned by the Metal Oxide Frameworks is reported in green. Sample structures for the pillared (blue border) and disordered GNM (red border) are reported as insets.

3. Pillared Materials: State of the Art and Open Problems

In disordered nanoporous materials the porosity of the final structures depends on how the flakes interlace during all the phases of the preparation, which introduces a high degree of disorder and stochasticity. In order to reduce this issue and improve the control over the outcome, the idea rose of synthesizing layered structures separated by molecular pillars, i.e., organic molecules suitably designed with given lengths, rigidity and possibly other physic-chemical properties. The size of the pillars determines the inter-layer spacing, and controls the average pore size, together with the relative distance of the pillars on the sheets. The first realization of such structures traces back to almost two decades ago when, inspired by metal-organic frameworks chemistry, layered structures separated by diboronic acid molecules were first proposed [81]. The pillars adhesion exploited the reactivity of hydroxy groups of graphene oxide (GO) with the acidic groups, leading to GO frameworks (GOF). These were subsequently characterized via Xray diffraction and neutron scattering, and tested through their H_2 adsorption capability, whose low value indicates rather a small SSA value (hundreds of m^2/g). The synthesis procedure was recently optimized [34], obtaining values of SSA up to ~600 m^2/g and pores size up to 2 nm. At the same time, it was shown that in some cases, in polar solvents these material exhibits reversible swelling, posing doubts on the complete covalent nature of the layer linkage [35]. Using as pillars di-amine of different lengths [82] resulted in materials with tuneable interlayer distance in the range 0.8–1.1 nm, generally hydrophilic and insulating. In fact, ab initio calculations with simplified model systems demonstrated that achieving electric conductivity in these materials is not easy, due to the rupture of the aromaticity at the linkage sites [80].

The optimization of these materials depends on their use: for electric energy storage, both a finely tuned pore size and the conductivity are important. Therefore different synthesis routes were explored, involving reduction of GO, either *ex post* [83] or directly starting from reduced GO (rGO); in the latter case, most of the proposed reaction exploit the chemistry of diazonium salts radicals, selectively reacting with the defective sites of rGO [84–86]. Among the best performances in terms of SSA where obtained with a two-step procedure: the rGO was first functionalized with benzoic acid [85], obtaining a layered material with good porosity, but scarce conductivity. Polyaniline was subsequently synthesised in situ obtaining a composite material with larger average inter-layer distance though smaller average pore size, and with improved electric conductance. Alternative routes to tailor inter-layer distance and porosity involve cross-linking by aryl-aryl reaction of rGO functionalized with iodo-phenyls [86] or Zn+ coordination of rGO functionalized with azobenzoic acid [87]. A summary of the recent literature on experimental and theoretical structural determinations of these materials is reported in Table 2. The main structural characteristics are also reported in Figure 1.

Table 2. Pillared materials derived by Graphite Oxide (GO) or reduced GO (rGO) flakes and their structural characteristics.

Precursor	Pillars	Reaction/Method	SSA m^2/g	Structural Features	H_2 Uptake (% at 77K)	Ref.
GO	Diboronic acid	Solvothermal Acid+OH dehydration	~200	~11 Å interlayer spacing; pillars distance: 7–8 Å	1% experiment 5% simulation	2010 [80]
GO	Diboronic acid	Solvothermal	500–600	Interlayer: 8–15 (swelling) Pore size > 2 nm	~1.5%	2015 [34]
GO	"tetrapod" amine	Solvothermal	>660	Interlayer: 10–13 to ~16 Å (swelling)	~1.5%	2017 [35]
GO	Different types of diamine	Cross-linking, thermally promoted		Interlayer 8.5–11 Å Pillar dist ~10 Å		2019 [84]
GO reduced	1-6 diaminohexane	Cross-linking	150–200	Inter layer: 7.8 Å Pore size: 1 nm, 15 nm		2018 [85]
rGO	Aryl bis-diazonium salts (and variants)	Radical reaction	200–400	Interlayer: 5–10 Å inter-pillar ext: ~5 Å		2016 [86]
rGO	Benzoic acid, polyaniline	Polyaniline is grown on benzoic acid on flakes	330	Inter layer 1.5–2.5 nm Density 0.68 g/cm^3 Pore size 0.8 nm		2015 [87]
rGO	4-iodophenyl diazonium salts	Aryl-aryl coupling reaction for cross-linking		Pore size 1–10 nm		2015 [88]
rGO	Azobenzoic acid-based ligands	Zn^{2+} coordination for cross-linking		inter-layer distance ~3 nm in the hydrogel		2012 [89]
graph	Diboronic acid variants	Density Functional Theory, Tight binding		Interlayer 1.1–2.2 nm inter pillar 3–5 Å	1.5%	2019 [84]
graph	nanotubes	Density Functional Theory, Grand Canonical MC		1.2 nm interlayer, 1.5 nm inter-pillar	6%	2017 [71,72]
GO, gr	Organic aromatic pillars	Reax FF Grand Canonical MC		Pore size 0.8,1,1.1 nm Inter layer ~3 nm	~4%	2017 [73,74]

Although steps forward have been done in the control of the functionalization, the performances of these materials are not better than those of disordered GNM: SSA is at best several hundreds, far below the theoretical limits and below the simulation predictions. In fact, both the carbon-only model systems including nanotube-pillars [88,89] and the molecular-pillared model systems [81,90], display, in simulations, GD uptake (and SSA) 5–10 times larger than the measured ones, besides the theoretical capability of efficient gas sorting [91], desalination [92], and interesting mechanical properties [93].

Although the origin of the theory-experiment discrepancy is not clear, it was shown that ideal structures with nearly flat sheets and regularly spaced pillars display a better performances in simulations [84], and that the adsorption performances depend on the fine tuning of the pillars distance, which must be large enough to allow the molecules access and hosting in a layer on the surface, but not too large, in order to maximize the GD. Therefore, a regular and controlled patterning seems the key for obtaining highly performing pillared materials.

4. Multilayers from Epitaxy: A Perspective

The reason why the pillars distribution is poorly controllable is encoded in the use of GO (or rGO) flakes as precursors: the covalent bonding of pillars or anchors exploits the presence of epoxy/hydroxy groups or defects, which are reactive sites [94]. However, these are randomly distributed, and their concentration is variable, and not easily tailorable [95]. In addition, flakes edges are also very reactive, attracting a relatively large number of functional groups, which introduces further disorder in the structure. Finally, the environmental conditions that promote the reaction (temperature, solvent, etc.) can favor aggregation in an almost unpredictable way. From this point of view, using epitaxial graphene as precursor would in principle bring some advantage, mainly related to the regularity of the material and to its laying on an extended solid support. In fact, this would allow a direct control over functionalization and check of results e.g., with atomic resolved microscopy techniques. One popular technique to produce supported graphene is chemical vapor deposition of carbon-rich compounds over metal substrates (after their cracking) [96]. Alternatively, one can use carbon rich substrates, such as SiC, and let the carbon layers reconstruct in the honeycomb lattice by selective evaporation of Si from surfaces with specific symmetries [97]. In both cases, one obtains macroscopic almost defectless single layers. In general, perfect graphene is poorly reactive, because of its fully delocalized stable sp^2 electronic system. Clearly, reactivity can be brought back by reintroducing defects, e.g., by nitrogen sputtering [98], which creates either substitutional or structural defects, proven to act as seeds for adhesion of metal clusters or hydrogen [99]. However, these defects are introduced randomly, pushing back to the same problems as in GO flakes.

Indeed, specific kinds of epitaxial graphene offer different possibilities, which exploit the interaction with the substrate. For instance, radicals of diazonium salts are able to attach to sp^2 sites but manifests a preference for graphene on hydrophilic substrates [100], due to charge accumulation effects. A similar effect is observed for graphene on metals such as iridium or ruthenium [101,102], where, in addition, a spatially modulated reactivity is created following the nano-metric moiré pattern of corrugation. This open the road to substrate driven functionalization, with the possibility of creating chemical nano-patterns following the symmetry of the moiré superlattice. Similar effects were obtained by intercalating metal clusters in between graphene and an insulating substrate [103], where the preferential adhesion of the radicals was observed in proximity of the metal cluster. The enhancement of reactivity (towards aryl radicals) is also observed on non-metallic substrates, such as patterned SiO_2 [104] and on the protruding areas of the natural moiré corrugation lattice of monolayer graphene on SiC (towards atomic hydrogen [26]). In these cases, it is attributed to the curvature [24]. In fact, both rippling and strain [105] produce charge inhomogeneities. Therefore, supported graphenes with moiré patterns are very promising materials for substrate driven regular nano-patterning.

We now focus on graphene on SiC, to further explore this concept. It is important to observe that graphene on SiC is not a single material but includes different types of 2D carbons [106] that can be obtained with different procedures (see Figure 2a). Upon Si evaporation from the Si-rich surfaces with hexagonal symmetry, excess carbon produces in the first instance a hexagonal carbon buffer layer (BL) [107], covalently bound to the substrate, and partially sp^3 hybridized. The bonds and corrugations follow a moiré pattern, due to the mismatch of the two lattices, displaying a hexagonal super-lattice of ~3.2 nm side, made of sharp crests and peaks with sp^3-like pyramidal configuration [15]. Fully sp^2 graphene can be obtained continuing evaporation: another BL forms under the first one, which is detached and becomes the so-called Mono-Layer graphene (MLG), characterised by a corrugation pattern with the same symmetry as BL, though smother [108]. Alternatively, the BL can be detached by intercalating H [108] or metals [109], obtaining the Quasi-Free-Standing Monolayer graphene (QFMLG) [110]. This is ideally flat, but displays in reality localized concavities, occupying the sites of a lattice roughly corresponding to 6 × 6 of SiC [111] with ~1.8 nm side, which were associated with vacancies of H in the intercalating layer [112]. The electronic structure is strongly affected by these defects, since Si dangling bonds produce electronic states localized near the Fermi level [113].

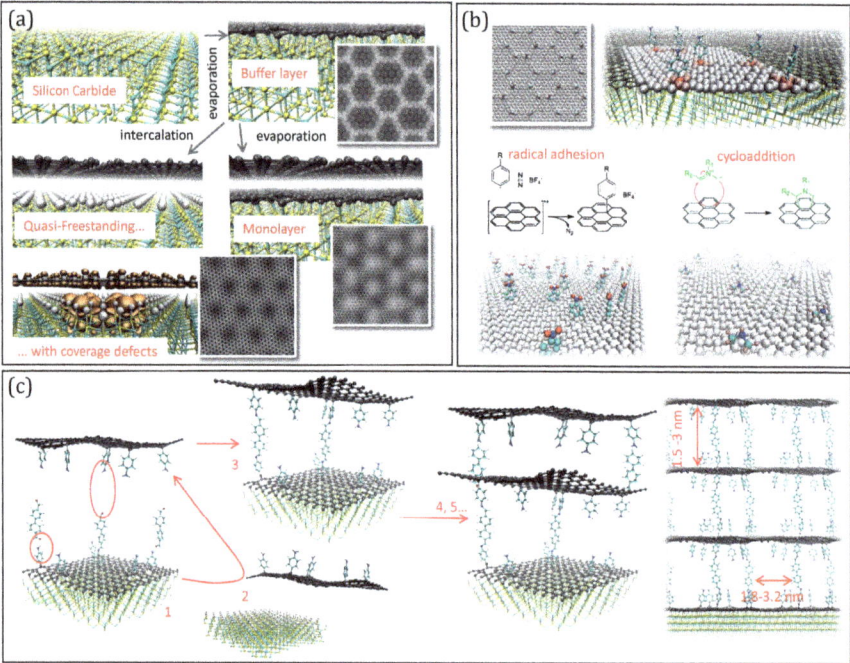

Figure 2. (**a**) Summary of the production of graphenes on SiC: the Buffer layer (BL by evaporation of Si), graphene monolayer (GML by subsequent evaporation) and quasi-free-standing monolayer (QFMLG, by intercalation of H or metal). The simulated Scanning Tunnel Microscopy images are reported for the BL, for the ML and for the QFSML with defects in the intercalation coverage layer. (**b**) Scheme of possible functionalization reactions exploiting the corrugation pattern of the BL. (**c**) Scheme of a possible strategy to build pillared multilayers: after pillaring (1), the cross-linking should occur with a previously detached functionalized sheet (2 to 3), and be re-iterated (4, 5 . . .) to give a regular structure.

While all of the different carbon layers on SiC display charge inhomogeneities following a regular nano-pattern induced by the interaction with the substrate, either mediated by the hybridization, by the corrugation or by the vacancies in the intercalation coverage, only the GML was tested on its reactivity, showing selective H adhesion on the crests [26]. On the other hand, the localised electronic states forming on the QFMLG in corresponding of H-vacancies have various sizes and shapes, depending on the number and relative location of vacant sites and their energy is organized in groups of levels near the Fermi energy [113], indicating a possible propensity to electrophile attack. Even more interesting from the functionalization is the buffer layer (Figure 2b) since it displays the strongest deviation from graphene symmetry and the sharper definition of the moiré pattern [15]. Specifically, the sp^3 cusps at the vertices of the moiré super-lattice are likely to be highly reactive sites in general, not only towards radicals, but possibly also towards e.g., dissociative chemisorption of H_2. Conversely, the protruding crests, organized in diene like structures, and the intruding areas, organized in "benzene-like" rings [15] are likely to be attractive for cyclo-addition reactions [114], leading to a spatially complementary selectivity.

Clearly, the BL functionalization should be viewed as the first, yet fundamental, step of a procedure involving the multilayer formation (see Figure 2c): once the molecular anchors are attached, pillars of different length can be added exploiting, e.g., solvothermal de-hydration reactions; subsequently the layer should be exposed to similarly functionalised layers (previously detached by the substrate by intercalation) which have to be stacked and cross-linked. These steps are also taken in the already realized synthesis of pillared materials from GO or rGO flakes [90–99]. However, using regularly patterned precursors would offer two unique advantages: first, the space matching of cross-linking groups can potentially trigger the self-assembly of the sheets, greatly impring the efficiency of the process, and second, the final result would be a structure with pillars at controlled distance in the range of 2–3 nm. This, together with the inter-layer distance controlled by the pillar length, will result in a structure with pre-determined porosity. Clearly, exploring experimentally this strategy would benefit of preliminary computer simulations, which are currently work in progress.

5. Summary, Conclusions and Possible Developments

In summary, we have reported three possible routes to produce graphene-based materials with porosity on the nano-scale, ordered by increasing capability of control and tailoring of the final structure. The first class produces the disordered nanoporous scaffolds from GO or rGO flakes. These can reach large values of SSA and are, up to now, the most interesting for gas storage. However, controlling their final structure is not straightforward, because of the disordered structure of the precursors and of the stochastic nature of their combination during the production procedure. With the aim of controlling at least a part of the variables determining the porosity, the second strategy introduces on the flakes pillars molecules with pre-determined lengths and shapes. This produces a class of materials with average pore sizes at the nano-metric scale, matching with the size of electrolytes and therefore suitable for the use in electric and electro-chemical storage. However, the average value of the SSA of these materials is rather low, and the poor control over the distribution of the pillars on the sheet introduces disorder, preventing a full optimization, not only for supercapacitors and batteries, but also in catalysis and filtering applications.

A third route is currently in its infancy, which would provide a full control over the distribution and location of the pillars. This considers as precursors epitaxial graphene and exploits the electronic inhomogeneities of the sheet produced by the interaction with the substrate, typically following a nano-metric moiré pattern, for the controlled chemical functionalization. Although the first timid steps (selective functionalization with atoms or small molecules) were demonstrated, the way is long towards the production of multi-layers.

The support of computer modeling and simulations is essential in all cases: in the case of disordered scaffolds, the main issue is to create realistic models and to understand the relationship between production procedure and final structure, and between the latter and the adsorption performances; for the pillared (r) GO materials, the challenge is to control the concentration and location of the pillars and predict the properties as a function of the used pillar. Most of all, computer modeling will be of outmost importance in the pillared multi-layers building from epitaxial graphene. In this case, the simulation of the pillaring, stacking and cross-linking would be essential to give indications for the experimental realization of the procedure. Though extremely challenging, this strategy might give a full control over all the structural features of the resulting structure, and–acting on the nature of pillars–might allow to create brand new materials with tailored and unprecedented properties, such as locally tuned elasticity or conductivity, reactivity to external fields, optical response, and others.

Funding: This research was funded by EU-H2020, Graphene-Core1 (agreement No. 696656) and Core2 (agreement No. 75219), MCSA (agreement No. 657070), by CINECA awards IsB11_flexogra (2015), IsC36_ElMaGRe (2015), IsC44_QFSGvac (2016), IsC44_ReIMCGr (2016), IsC61_MGchpDFA (2018) IsC69_EFaRe (2019), IsB19_DiNaGra (2019) and PRACE "Tier0" award Pra13_2016143310 (2016).

Acknowledgments: We thank Stefan Heun, Paolo Giannozzi, Camilla Coletti, Yuya Murata, and Vittorio Pellegrini for useful discussions.

Conflicts of Interest: The authors declare no conflict of interest.

References

1. Geim, A.K.; Novoselov, K.S. The rise of graphene. *Nat. Mater.* **2007**, *6*, 183–191. [CrossRef] [PubMed]
2. Castro Neto, A.H.; Guinea, F.; NMRPeres Novoselov, K.S.; Geim, A.K. The electronic properties of graphene. *Rev. Mod. Phys.* **2009**, *81*, 109. [CrossRef]
3. Lee, C.; Wei, X.; Kysar, J.W.; Hone, J. Measurement of the Elastic Properties and Intrinsic Strength of Monolayer Graphene. *Science* **2008**, *321*, 385. [CrossRef] [PubMed]
4. Camiola, V.D.; Farchioni, R.; Pellegrini, V.; Tozzini, V. Hydrogen transport within graphene multilayers by means of flexural phonons. *2D Mater.* **2015**, *2*, 014009. [CrossRef]
5. Fasolino, A.; Los, J.H.; Katsnelson, M.I. Intrinsic ripples in graphene. *Nat. Mater.* **2007**, *6*, 858–861. [CrossRef] [PubMed]
6. Bonaccorso, F.; Colombo, L.; Yu, G.; Stoller, M.; Tozzini, V.; Ferrari, A.C.; Ruoff, R.S.; Pellegrini, V. Graphene, related two-dimensional crystals, and hybrid systems for energy conversion and storage. *Science* **2015**, *347*, 1246501. [CrossRef]
7. Sun, P.; Wang, K.; Zhu, H. Recent Developments in Graphene-Based Membranes: Structure, Mass-Transport Mechanism and Potential Applications. *Adv. Mater.* **2016**, *28*, 2287–2310. [CrossRef]
8. Büch, H.; Rossi, A.; Forti, S.; Convertino, D.; Tozzini, V.; Coletti, C. Superlubricity of epitaxial monolayer WS2 on graphene. *Nano Res.* **2018**, *11*, 5946–5956. [CrossRef]
9. Chee, W.K.; Lim, H.N.; Zainal, Z.; Huang, N.M.; Harrison, I.; Andou, Y. Flexible Graphene-Based Supercapacitors: A Review. *J. Phys. Chem. C* **2016**, *120*, 4153–4172. [CrossRef]
10. Mahmoudi, T.; Wang, Y.; Hahn, Y.-B. Graphene and its derivatives for solar cells application. *Nano Energy* **2018**, *47*, 51–65. [CrossRef]
11. Denis, P.A.; Huelmo, C.P. Martins AS Band Gap Opening in Dual-Doped Monolayer Graphene. *J. Phys. Chem. C* **2016**, *120*, 13. [CrossRef]
12. Iyakutti, K.; Kumar, E.M.; Lakshmi, I.; Thapa, R.; Rajeswarapalanichamy, R.; Surya, V.J.; Kawazoe, Y. Effect of surface doping on the band structure of graphene: A DFT study. *J. Mater. Sci.* **2016**, *27*, 2728–2740. [CrossRef]
13. Deng, S.; Berry, V. Wrinkled, rippled and crumpled graphene: An overview of formation mechanism, electronic properties, and applications. *Mater. Today* **2016**, *19*, 197–212. [CrossRef]
14. Zaminpayma, E.; Emami Razavi, M.; Nayebi, P. Electronic properties of graphene with single vacancy and Stone-Wales defects. *Appl. Surf. Sci.* **2017**, *31*, 101–106. [CrossRef]
15. Nair, M.N.; Palacio, I.; Celis, A.; Zobelli, A.; Gloter, A.; Kubsky, S.; Turmaud, J.-P.; Conrad, M.; Berger, C.; de Heer, W.; et al. Band Gap Opening Induced by the Structural Periodicity in Epitaxial Graphene Buffer Layer. *Nano Lett.* **2017**, *174*, 2681–2689. [CrossRef]
16. Cavallucci, T.; Tozzini, V. Intrinsic structural and electronic properties of the Buffer Layer on Silicon Carbide unraveled by Density Functional Theory. *Sci. Rep.* **2018**, *8*, 13097. [CrossRef]
17. Loh, K.P.; Tong, S.W.; Wu, J. Graphene and Graphene-like Molecules: Prospects in Solar Cells. *J. Am. Chem. Soc.* **2016**, *138*, 1095–1102. [CrossRef]
18. Cohen-Tanugi, D.; Lin, D.-C.; Grossman, J.C. Multilayer Nanoporous Graphene Membranes for Water Desalination. *Nano Lett.* **2016**, *16*, 1027–1033. [CrossRef]
19. Guinea, F.; Katsnelson, M.I.; Geim, A.K. Energy gaps and a zero-field quantum Hall effect in graphene by strain engineering. *Nat. Phys.* **2010**, *6*, 30. [CrossRef]
20. Hicks, J.; Tejeda, A.; Taleb-Ibrahimi, A.; Nevius, M.S.; Wang, F.; Shepperd, K.; Palmer, J.; Bertran, F.; Le Fèvre, P.; Kunc, J.; et al. A wide-bandgap metal–semiconductor–metal nanostructure made entirely from graphene. *Nat. Phys.* **2012**, *9*, 49. [CrossRef]
21. Rossi, A.; Piccinin, S.; Pellegrini, V.; de Gironcoli, S.; Tozzini, V. Nano-Scale Corrugations in Graphene: A Density Functional Theory study of Structure, Electronic Properties and Hydrogenation. *J. Phys. Chem. C* **2015**, *119*, 7900. [CrossRef]
22. McKay, H.; Wales, D.J.; Jenkins, S.J.; Verges, J.A.; de Andres, P.L. Hydrogen on graphene under stress: Molecular dissociation and gap opening. *Phys. Rev. B* **2010**, *81*, 075425. [CrossRef]

23. Goler, S.; Coletti, C.; Tozzini, V.; Piazza, V.; Mashoff, T.; Beltram, F.; Pellegrini, V.; Heun, S. Influence of Graphene Curvature on Hydrogen Adsorption: Towards Hydrogen Storage Devices. *J. Phys. Chem. C* **2013**, *117*, 11506. [CrossRef]
24. Boukhvalov, D.W.; Son, Y.-W. Covalent Functionalization of Strained Graphene. *Chem. Phys. Chem.* **2012**, *13*, 1463. [CrossRef]
25. Wang, Z.F.; Zhang, Y.; Liu, F. Formation of hydrogenated graphene nanoripples by strain engineering and directed surface self-assembly. *Phys. Rev. B* **2011**, *83*, 041403(R). [CrossRef]
26. Cavallucci, T.; Kakhiani, K.; Farchioni, R.; Tozzini, V. Morphing Graphene-Based Systems for Applications: Perspectives from Simulations. In *GraphITA Carbon Nanostructures*; Springer: Cham, Switzerland, 2017; pp. 87–111.
27. Camiola, V.D.; Farchioni, R.; Cavallucci, T.; Rossi, A.; Pellegrini, V.; Tozzini, V. Hydrogen storage in rippled graphene: Perspectives from multi-scale simulations. *Front. Mater.* **2015**, *2*, 3. [CrossRef]
28. Quesnel, E.; Roux, F.; Emieux, F.; Faucherand, P.; Kymakis, E.; Volonakis, G.; Giustino, F.; Martín-García, B.; Moreels, I.; Gürsel, S.A.; et al. Graphene-based technologies for energy applications, challenges and perspectives. *2D Mater.* **2015**, *2*, 030204. [CrossRef]
29. Klechikov, A.G.; Mercier, G.; Merino, P.; Blanco, S.; Merino, C.; Talyzin, A.V. Hydrogen storage in bulk graphene-related materials. *Micropor. Mesopor. Mater.* **2015**, *210*, 46–51. [CrossRef]
30. Qiu, B.; Xing, M.; Zhang, J. Recent advances in three-dimensional graphene based materials for catalysis applications. *Chem. Soc. Rev.* **2018**, *47*, 2165. [CrossRef]
31. Bustillos, J.; Zhang, C.; Boesl, B.; Agarwal, A. Three-Dimensional Graphene Foam–Polymer Composite with Superior Deicing Efficiency and Strength. *ACS Appl. Mater. Interfaces* **2018**, *10*, 5022–5029. [CrossRef]
32. Neves, A.I.S.; Rodrigues, D.P.; De Sanctis, A.; Alonso, E.T.; Pereira, M.S.; Amaral, V.S.; Melo, L.V.; Russo, S.; de Schrijver, I.; HAlves, M.F. Craciun Towards conductive textiles: Coating polymeric fibres with graphene. *Sci. Rep.* **2017**, *7*, 4250. [CrossRef] [PubMed]
33. Zheng, Z.; Liu, Y.; Bai, Y.; Zhang, J.; Han, Z.; Ren, L. Fabrication of biomimetic hydrophobic patterned graphene surface with ecofriendly anti-corrosion properties for Al alloy. *Coll. Surf. A Physicochem. Eng. Asp.* **2016**, *500*, 64–71. [CrossRef]
34. Xu, Y.; Sheng, K.; Li, C.; Shi, G. Self-assembled graphene hydrogel via a one-step hydrothermal process. *ACS Nano* **2010**, *4*, 4324. [CrossRef]
35. Mercier, G.; Klechikov, A.; Hedenstroöm, M.; Johnels, D.; Baburin, I.A.; Seifert, G.; Mysyk, R.; Talyzin, A.V. Porous Graphene Oxide/Diboronic Acid Materials: Structure and Hydrogen Sorption. *J. Phys. Chem. C* **2015**, *119*, 27179–27191. [CrossRef]
36. Sun, J.; Morales-Lara, F.; Klechikov, A.; Talyzin, A.V.; Baburin, A.; Seifert, G.; Cardano, F.; Baldrighi, M.; Frasconi, M.; Giordani, S. Porous graphite oxide pillared with tetrapod-shaped molecules. *Carbon* **2017**, *120*, 145–156. [CrossRef]
37. Liang, C.; Li, Z.; Dai, S. Mesoporous carbon materials: Synthesis and modification. *Angew. Chem. Int. Ed.* **2008**, *47*, 3696–3717. [CrossRef] [PubMed]
38. Talyzin, A.V.; Mercier, G.; Klechikov, A.; Hedenström, M.; Johnels, D.; Wei, D.; Cotton, D.; Moons, A.E. Brodie vs Hummers graphite oxides for preparation of multi-layered materials. *Carbon* **2017**, *115*, 430–440. [CrossRef]
39. Talyzin, A.V.; Szabó, T.; Dékány, I.; Langenhorst, F.; Sokolov, P.S.; Solozhenko, V.L. Nanocarbons by High-Temperature Decomposition of Graphite Oxide at Various Pressures. *J. Phys. Chem. C* **2009**, *113*, 11279–11284. [CrossRef]
40. Zhu, Y.; Murali, S.; Stoller, M.D.; Ganesh, K.J.; Cai, W.; Ferreira, P.J.; Pirkle, A.; Wallace, R.M.; Cychosz, K.A.; Thommes, M.; et al. Carbon-based supercapacitors produced by activation of graphene. *Science* **2011**, *332*, 1537–1541. [CrossRef]
41. Zhu, Y.; Murali, S.; Stoller, M.D.; Velamakanni, A.; Piner, R.D.; Ruoff, R.S. Microwave assisted exfoliation and reduction of graphite oxide for ultracapacitors. *Carbon* **2010**, *48*, 2118–2122. [CrossRef]
42. Zhang, C.; Lv, W.; Xie, X.; Tang, D.; Liu, C.; Yang, Q.-H. Towards low temperature thermal exfoliation of graphite oxide for graphene production. *Carbon* **2013**, *62*, 11–24. [CrossRef]
43. Kovtun, A.; Treossi, E.; Mirotta, N.; Scidà, A.; Liscio, A.; Christian, M.; Valorosi, F.; Boschi, A.; Young, R.J.; Galiotis, C.; et al. Benchmarking of graphene-based materials: Real commercial products versus ideal graphene. *2D Mater.* **2019**, *6*, 025006. [CrossRef]

44. Klechikov, A.; Mercier, G.; Sharifi, T.; Baburin, I.A.; Seifert, G.; Talyzin, A.V. Hydrogen storage in high surface area graphene scaffolds. *Chem. Comm.* **2015**, *51*, 15280–15283. [CrossRef] [PubMed]
45. Raccichini, R.; Varzi, A.; Passerini, S.; Scrosati, B. The role of graphene for electrochemical energy storage. *Nature Mater.* **2015**, *14*, 271–279. [CrossRef] [PubMed]
46. Chen, K.; Song, S.; Li, F.; Xue, D. Structural design of graphene for use in electrochemical energy storage devices. *Chem. Soc. Rev.* **2015**, *44*, 6230–6257. [CrossRef] [PubMed]
47. Minuto, F.D.; Policicchio, A.; Aloise, A.; Agostino, R.G. Liquid-like hydrogen in the micropores of commercial activated carbons. *Int. J. Hydrog. Energy* **2015**, *40*, 14562–14572. [CrossRef]
48. Nomura, K.; Nishihara, H.; Kobayashi, N.; Asada, T.; Kyotani, T. 4.4 V supercapacitors based on super-stable mesoporous carbon sheet made of edge-free graphene walls. *Energy Environ. Sci.* **2019**, *12*, 1542–1549. [CrossRef]
49. Tsai, W.-Y.; Lin, R.; Murali, S.; Zhang, L.; McDonough, J.K.; Ruoff, R.S.; Taberna, P.-L.; Gogotsi Yu Simon, P. Outstanding performance of activated graphene based supercapacitors in ionic liquid electrolyte from −50 to 80 °C. *Nano Energy* **2013**, *2*, 403–411. [CrossRef]
50. Méndez-Morales, T.; Ganfoud, N.; Li, Z.; Haefele, M.; Rotenberg, B.; Salanne, M. Performance of microporous carbon electrodes for supercapacitors: Comparing graphene with disordered materials. *Energy Storage Mater.* **2019**, *17*, 88–92. [CrossRef]
51. Kondrat, S.; Kornyshev, A.A. Pressing a spring: What does it take to maximize the energy storage in nanoporous supercapacitors? *Nanoscale Horiz.* **2016**, *1*, 45–52. [CrossRef]
52. Salanne, M.; Rotenberg, B.; Naoi, K.; Kaneko, K.; Taberna, P.-L.; Grey, C.P.; Dunn, B.; Simon, P. Efficient storage mechanisms for building better supercapacitors. *Nat. Energy* **2016**, *1*, 16070. [CrossRef]
53. Huang, J.-Q.; Zhuang, T.-Z.; Zhang, Q.; Peng, H.-J.; Chen, C.-M.; Wei, F. Permselective Graphene Oxide Membrane for Highly Stable and Anti-Self-Discharge Lithium–Sulfur Batteries. *ACS Nano* **2015**, *9*, 3002–3011. [CrossRef] [PubMed]
54. Sun, Y.; Tang, J.; Zhang, K.; Yuan, J.; Li, J.; Zhu, D.-M.; Ozawa, K.; Qin, L.-C. Comparison of reduction products from graphite oxide and graphene oxide for anode applications in lithium-ion batteries and sodium-ion batteries. *Nanoscale* **2017**, *9*, 2585–2595. [CrossRef] [PubMed]
55. Bellucci, L.; Tozzini, V. *In Silico Design of Nano-Porous Graphene Scaffolds*, in preparation.
56. Yang, T.; Lin, H.; Zheng, X.; Loh, K.P.; Jia, B. Tailoring pores in graphene-based materials: From generation to applications. *J. Mater. Chem. A* **2017**, *5*, 16537–16558. [CrossRef]
57. Fang, T.-H.; Lee, Z.-W.; Chang, W.-J.; Huang, C.-C. Determining porosity effect on the thermal conductivity of single-layer graphene using a molecular dynamics simulation. *Phys. E Low Dimens. Syst. Nanostruct.* **2019**, *106*, 90–94. [CrossRef]
58. Wu, C.D.; Fang, T.H.; Lo, J.Y.; Feng, Y.L. Molecular dynamics simulations of hydrogen storage capacity of few-layer graphene. *J. Mol. Model.* **2013**, *19*, 3813–3819. [CrossRef]
59. Gotzias, A.; Tylianakis, E.; Froudakis, G.; Steriotis, T.H. Theoretical study of hydrogen adsorption in oxygen functionalized carbon slit pores. *Micropor. Mesopor. Mater.* **2012**, *154*, 38–44. [CrossRef]
60. Cabria, I.; López, M.J.; Alonso, J.A. The optimum average nanopore size for hydrogen storage in carbon nanoporous materials. *Carbon* **2007**, *45*, 2649–2658. [CrossRef]
61. Georgakis, M.; Stavropoulos, G.; Sakellaropoulos, G.P. Alteration of graphene based slit pores and the effect on hydrogen molecular adsorption: A simulation study. *Micropor. Mesopor. Mater.* **2014**, *191*, 67–73. [CrossRef]
62. Kowalczyk, P.; Gauden, P.A.; Furmaniak, S.; Terzyk, A.P.; Wisniewski, M.; Ilnicka, A.; Łukaszewicz, J.; Burian, A.; Włoch, J.; Neimark, A.V. Morphologically disordered pore model for characterization of micro-mesoporous carbons. *Carbon* **2017**, *111*, 358–370. [CrossRef]
63. Patchkovskii, S.; John, S.T.; Yurchenko, S.N.; Zhechkov, L.; Heine, T.; Seifert, G. Graphene nanostructures as tunable storage media for molecular hydrogen. *Proc. Natl. Acad. Sci. USA* **2005**, *102*, 10439–10444. [CrossRef] [PubMed]
64. Kuchta, B.; Firlej, L.; Mohammadhosseini, A.; Boulet, P.; Beckner, M.; Romanos, J.; Pfeifer, P. Hypothetical high-surface-area carbons with exceptional hydrogen storage capacities: Open carbon frameworks. *J. Am. Chem. Soc.* **2012**, *134*, 15130–15137. [CrossRef] [PubMed]
65. Krainyukova, N.V.; Zubarev, E.N. Carbon honeycomb high capacity storage for gaseous and liquid species. *Phys. Rev. Lett.* **2016**, *116*, 055501. [CrossRef] [PubMed]

66. Li, L.; Xu, M.; Song, W.; Ovcharenko, A.; Zhang, G.; Jia, D. The effect of empirical potential functions on modeling of amorphous carbon using molecular dynamics method. *Appl. Surf. Sci.* **2013**, *286*, 287–297. [CrossRef]
67. Tersoff, J. Modelling solid-state chemistry: Interatomic potentials for multicomponent systems. *Phys. Rev.* **1989**, *39*, 5566–5568. [CrossRef]
68. Stuart, S.J.; Tutein, A.B.; Harrison, J.A. A reactive potential for hydrocarbons with intermolecular interactions. *J. Chem. Phys.* **2000**, *112*, 6472–6648. [CrossRef]
69. Ghiringhelli, L.M.; Valeriani, C.; Los, J.H.; Meijer, E.J.; Fasolino, A.; Frenkel, D. State-of-the-art models for the phase diagram of carbon and diamond nucleation. *Mol. Phys.* **2010**, *106*, 2011–2038. [CrossRef]
70. Chenoweth, K.; Van Duin, A.C.; Goddard, W.A. ReaxFF reactive force field for molecular dynamics simulations of hydrocarbon oxidation. *J. Phys. Chem. A* **2008**, *112*, 1040–1053. [CrossRef]
71. Ganfoud, N.; Sene, A.; Haefele, M.; Marin-Laflèche, A.; Daffos, B.; Taberna, P.L.; Salanne, M.; Simon, P.; Rotenberg, B. Effect of the carbon microporous structure on the capacitance of aqueous supercapacitors Energy Storage Mater. *Energy Storage Mater* **2019**, *21*, 190–195. [CrossRef]
72. Mejia-Mendoza, L.M.; Valdez-Gonzalez, M.; Muniz, J.; Santiago, U.; Cuentas-Gallegos, A.K.; Robles, M. A theoretical approach to the nanoporous phase diagram of carbon. *Carbon* **2017**, *120*, 233–0243. [CrossRef]
73. Thompson, M.W.; Dyatkin, B.; Wang, H.-W.; Turner, C.H.; Sang, X.; Unocic, R.R.; Iacovella, C.R.; Gogotsi, Y.; van Duin Cummings, P.T. An Atomistic Carbide-Derived Carbon Model Generated Using ReaxFF-Based Quenched Molecular Dynamics. *J. Carbon Res. C* **2017**, *3*, 32. [CrossRef]
74. Ranganathan, R.; Rokkam, S.; Desa, T.; Keblinski, P. Generation of amorphous carbon models using liquid quench method: A reactive molecular dynamics study. *Carbon* **2017**, *113*, 87–99. [CrossRef]
75. Surendra, K.J.; Roland, J.-M.P.; Pikunic, J.P.; Gubbins, K.E. Molecular Modeling of Porous Carbons Using the Hybrid Reverse Monte Carlo Method. *Langmuir* **2006**, *22*, 9942–9948.
76. Farmahini, A.H.; Bhatia, S.K. Hybrid Reverse Monte Carlo simulation of amorphous carbon: Distinguishing between competing structures obtained using different modeling protocols. *Carbon* **2015**, *83*, 53–70. [CrossRef]
77. Sarkisov, L. Accessible Surface Area of Porous Materials: Understanding Theoretical Limits. *Adv. Mater.* **2012**, *24*, 3130–3133. [CrossRef]
78. Baburin, I.A.; Klechikov, A.; Mercier, G.; Talyzin, A.; Seifert, G. Hydrogen adsorption by perforated graphene. *Int. J. Hydrog. Energy* **2015**, *40*, 6594–6599. [CrossRef]
79. Kostoglou, N.; Tarat, A.; Walters, I.; Ryzhkov, V.; Tampaxis, C.; Charalambopoulou, G.; Steriotis, T.; Mitterer, C.; Rebholz, C. Few-layer graphene-like flakes derived by plasma treatment: A potential material for hydrogen adsorption and storage. *Microporous Mesoporous Mater.* **2016**, *225*, 482–487. [CrossRef]
80. Klontzas, E.; Tylianakis, E.; Varshney, V.; Roy, A.K.; Froudakis, G.E. Organically interconnected graphene flakes: A flexible 3-D material with tunable electronic bandgap. *Sci. Rep.* **2019**, *9*, 13676. [CrossRef]
81. Burress, J.W.; Gadipelli, S.; Ford, J.; Simmons, J.M.; Zhou, W.; Yildirim, T. Graphene Oxide Framework Materials: Theoretical Predictions and Experimental Results. *Angew. Chem. Int. Ed.* **2010**, *49*, 8902–8904. [CrossRef]
82. Hung, W.-S.; Tsou, C.-H.; De Guzman, M.; An, Q.-F.; Liu, Y.-L.; Zhang, Y.a.-M.; Hu, C.-C.; Lee, K.-R.; Lai, J.-Y. Cross-Linking with Diamine Monomers To Prepare Composite Graphene Oxide-Framework Membranes with Varying d-Spacing. *Chem. Mater.* **2014**, *26*, 2983–2990. [CrossRef]
83. Banda, H.; Périé, S.; Daffos, B.; Taberna, P.-L.; Dubois, L.; Crosnier, O.; Simon, P.; Lee, D.; De Paëpe, G.; Duclairoir, F. Sparsely Pillared Graphene Materials for High- Performance Supercapacitors: Improving Ion Transport and Storage Capacity. *ACS Nano* **2019**, *13*, 1443–1453. [CrossRef] [PubMed]
84. Lee, K.; Yoon, Y.; Cho, Y.; Lee, S.M.; Shin, Y.; Lee, H.; Lee, H. Tunable Sub-nanopores of Graphene Flake Interlayers with Conductive Molecular Linkers for Supercapacitors. *ACS Nano* **2016**, *10*, 6799–6807. [CrossRef] [PubMed]
85. Sekar, P.; Anothumakkool, B.; Kurungot, S. 3D Polyaniline Porous Layer Anchored Pillared Graphene Sheets: Enhanced Interface Joined with High Conductivity for Better Charge Storage Applications. *ACS Appl. Mater. Interfaces* **2015**, *7*, 7661–7669. [CrossRef] [PubMed]
86. Yuan, K.; Xu, Y.; Uihlein, J.; Brunklaus, G.; Shi, L.; Heiderhoff, R.; Que, M.; Forster, M.; Chassé, T.; Pichler, T.; et al. Straightforward Generation of Pillared, Microporous Graphene Frameworks for Use in Supercapacitors. *Adv. Mater.* **2015**, *27*, 6714–6721. [CrossRef]

87. Lee, J.H.; Kang, S.; Jaworski, J.; Kwon, K.-Y.; Seo, M.L.; Lee, J.Y.; Jung, J.H. Fluorescent Composite Hydrogels of Metal–Organic Frameworks and Functionalized Graphene Oxide. *Chem. Eur. J.* **2012**, *18*, 765–769. [CrossRef]
88. Dimitrakakis, G.K.; Tylianakis, E.; Froudakis, G.E. Pillared Graphene: A New 3-D Network Nanostructure for Enhanced Hydrogen Storage. *Nano Lett.* **2008**, *8*, 3166. [CrossRef]
89. Hassani, A.; Taghi, M.; Mosavian, H.; Ahmadpour, A.; Farhadian, N. Hybrid molecular simulation of methane storage inside pillared graphene. *J. Chem. Phys.* **2015**, *142*, 234704. [CrossRef]
90. Pedrielli, A.; Taioli, S.; Garberoglio, G.; Pugno, N.M. Gas adsorption and dynamics in Pillared Graphene Frameworks. *Microporous Mesoporous Mater.* **2018**, *257*, 222–231. [CrossRef]
91. Garberoglio, G.; Pugno, N.M.; Taioli, S. Gas adsorption and separation in realistic and idealized frameworks of organic pillared graphene: A comparative study. *J. Phys. Chem. C* **2015**, *119*, 1980–1987. [CrossRef]
92. Mahdizadeh, S.J.; Goharshadi, E.K.; Akhlamadia, G. Seawater desalination using pillared graphene as a novel nano-membrane in reverse osmosis process: Nonequilibrium MD simulation study. *Phys. Chem. Chem. Phys.* **2018**, *20*, 22241. [CrossRef]
93. Wang, Y.C.; Zhu, Y.; FCWang Liu, X.Y.; Wu, H.A. Super-elasticity and deformation mechanism of three-dimensional pillared graphene network structures. *Carbon* **2017**, *118*, 588–596. [CrossRef]
94. Ciammaruchi, L.; Bellucci, L.; Comeron Castillo, G.; Martínez-DenegriSanchez, G.; Liu, Q.; Tozzini Martorell, J. Water splitting for hydrogen chemisorption in graphene oxide dynamically evolving to a graphane character lattice. *Carbon* **2019**, *153*, 234–241. [CrossRef]
95. Morimoto, N.; Kubo, T.; Nishina, Y. Tailoring the Oxygen Content of Graphite and Reduced Graphene Oxide for Specific Applications. *Sci. Rep.* **2016**, *6*, 21715. [CrossRef] [PubMed]
96. Zhang, Y.; Zhang, L.; Zhou, C. Review of Chemical Vapor Deposition of Graphene and Related Applications. *Acc. Chem. Res.* **2013**, *46*, 2329–2339. [CrossRef]
97. Riedl, C.; Coletti, C.; Starke, U. Structural and Electronic Properties of Epitaxial Graphene on SiC(0001): A Review of Growth, Characterization, Transfer Doping and Hydrogen Intercalation. *J. Phys. D Appl. Phys.* **2010**, *43*, 374009. [CrossRef]
98. Mashoff, T.; Convertino, D.; Miseikis, V.; Coletti, C.; Piazza, V.; Tozzini, V.; Beltram, F.; Heun, S. Increasing the active surface of titanium islands on graphene by nitrogen sputtering. *Appl. Phys. Lett.* **2015**, *106*, 083901. [CrossRef]
99. Takahashi, K.; Isobe, S.; Omori, K.; Mashoff, T.; Convertino, D.; Miseikis, V.; Coletti, C.; Tozzini, V.; Heun, S. Revealing the Multi-Bonding State Between Hydrogen and Graphene-Supported Ti Clusters. *J. Phys. Chem. C* **2016**, *120*, 12974. [CrossRef]
100. Wang, Q.H.; Jin, Z.; Kim, K.K.; Hilmer, A.J.; Paulus, G.L.C.; Shih, C.-J.; Ham, M.-H.; Sanchez-Yamagishi, J.D.; Watanabe, K.; Taniguchi, T.; et al. Understanding and controlling the substrate effect on graphene electron-transfer chemistry via reactivity imprint lithography. *Nat. Chem.* **2012**, *4*, 724. [CrossRef]
101. Navarro, J.J.; Leret, S.; Calleja, F.; Stradi, D.; Black, A.; Bernardo-Gavito, R.; Garnica, M. Organic Covalent Patterning of Nanostructured Graphene with Selectivity at the Atomic Level. *Nano Lett.* **2016**, *16*, 355–361. [CrossRef]
102. Romero-Muñiz, C.; Martín-Recio, A.; Pou, P.; Gómez-Rodríguez, J.M.; Pérez, R. Substrate-induced enhancement of the chemical reactivity in metal-supported graphene. *Phys. Chem. Chem. Phys.* **2018**, *20*, 19492–19499. [CrossRef]
103. Criado, A.; Melchionna, M.; Marchesan, S.; Prato, M. The Covalent Functionalization of Graphene on Substrates. *Angew. Chem.* **2015**, *54*, 10734. [CrossRef] [PubMed]
104. Wu, Q.; Wu, Y.; Hao, Y.; Geng, J.; Charlton, M.; Chen, S.; Ren, Y.; Ji, H.; Li, H.; Boukhvalov, D.W.; et al. Selective surface functionalization at regions of high local curvature in graphene. *Chem. Commun.* **2013**, *49*, 677–679. [CrossRef] [PubMed]
105. Bissett, M.A.; Konabe, S.; Okada, S.; Tsuji, M.; Ago, H. Enhanced chemical reactivity of graphene induced by mechanical strain. *ACS Nano* **2013**, *7*, 10335–10343. [CrossRef] [PubMed]
106. Bellucci, L.; Cavallucci, T.; Tozzini, V. From the Buffer Layer to Graphene on Silicon Carbide: Exploring Morphologies by Computer Modeling. *Front. Mater.* **2019**, *6*, 198. [CrossRef]
107. Goler, S.; Coletti, C.; Piazza, V.; Pingue, P.; Colangelo, F.; Pellegrini, V.; Emtsev, K.V.; Forti, S.; Starke, U.; Heun, S.; et al. Revealing the Atomic Structure of the Buffer Layer between SiC (0001) and Epitaxial Graphene. *Carbon* **2013**, *51*, 249. [CrossRef]

108. Cavallucci, T.; Tozzini, V. Multistable Rippling of Graphene on SiC: A Density Functional Theory Study. *J. Phys. Chem C* **2016**, *120*, 7670. [CrossRef]
109. Fiori, S.; Murata, Y.; Veronesi, S.; Rossi, A.; Coletti, C.; Heun, S. Li-intercalated graphene on SiC (0001): An STM study. *Phys. Rev. B* **2017**, *96*, 125429. [CrossRef]
110. Riedl, C.; Coletti, C.; Iwasaki, T.; Zakharov, A.A.; Starke, U. Quasi-Free-Standing Epitaxial Graphene on SiC Obtained by Hydrogen Intercalation. *Phys. Rev. Lett.* **2009**, *103*, 246804. [CrossRef]
111. Murata, Y.; Mashoff, T.; Takamura, M.; Tanabe, S.; Hibino, H.; Beltram, F.; Heun, S. Correlation between morphology and transport properties of quasi free standing monolayer graphene. *Appl. Phys. Lett.* **2014**, *105*, 221604. [CrossRef]
112. Murata, Y.; Cavallucci, T.; Tozzini, V.; Pavlíček, N.; Gross, L.; Meyer, G.; Takamura, M.; Hibino, H.; Beltram, F.; Heun, S. Atomic and electronic structure of Si dangling bonds in quasi-free-standing monolayer graphene. *Nano Res.* **2018**, *11*, 864. [CrossRef]
113. Cavallucci, T.; Murata, Y.; Heun, S.; Tozzini, V. Unraveling localized states in quasi free standing monolayer graphene by means of Density Functional Theory. *Carbon* **2018**, *130*, 466–474. [CrossRef]
114. Hess, L.H.; Lyuleeva, A.; Blaschke, B.M.; Sachsenhauser, M.; Seifert, M.; Garrido, J.A.; Coulombwall, A. Graphene Transistors with Multifunctional Polymer Brushes for Biosensing Applications. *ACS Appl. Mater. Interfaces* **2014**, *6*, 9705. [CrossRef] [PubMed]

© 2020 by the authors. Licensee MDPI, Basel, Switzerland. This article is an open access article distributed under the terms and conditions of the Creative Commons Attribution (CC BY) license (http://creativecommons.org/licenses/by/4.0/).

Review

Filled Carbon Nanotubes as Anode Materials for Lithium-Ion Batteries

Elisa Thauer [1], Alexander Ottmann [1], Philip Schneider [1], Lucas Möller [1], Lukas Deeg [1], Rouven Zeus [1], Florian Wilhelmi [1], Lucas Schlestein [1], Christoph Neef [1], Rasha Ghunaim [2,3], Markus Gellesch [2], Christian Nowka [2], Maik Scholz [2], Marcel Haft [2], Sabine Wurmehl [2,4], Karolina Wenelska [5], Ewa Mijowska [5], Aakanksha Kapoor [6], Ashna Bajpai [6], Silke Hampel [2] and Rüdiger Klingeler [1,7,*]

1. Kirchhoff Institute for Physics, Heidelberg University, INF 227, 69120 Heidelberg, Germany; elisa.thauer@kip.uni-heidelberg.de (E.T.); alex.ottmann@posteo.de (A.O.); schneider_philip@web.de (P.S.); lucas.moeller@me.com (L.M.); lukas-deeg@gmx.de (L.D.); rouven.zeus@gmx.net (R.Z.); florianwilhelmi@gmx.de (F.W.); lucasschlestein@gmx.de (L.S.); Christoph.Neef@isi.fraunhofer.de (C.N.)
2. Leibniz Institute for Solid State and Materials Research (IFW) Dresden, 01069 Dresden, Germany; rgonaim@ppu.edu (R.G.); M.Gellesch@bham.ac.uk (M.G.); c.nowka@ifw-dresden.de (C.N.); maik.scholz@ifw-dresden.de (M.S.); m.haft@ifw-dresden.de (M.H.); s.wurmehl@ifw-dresden.de (S.W.); s.hampel@ifw-dresden.de (S.H.)
3. Department of Applied Chemistry, Palestine Polytechnic University, Hebron P.O. Box 198, Palestine
4. Institute for Physics of Solids, Technical University of Dresden, 01062 Dresden, Germany
5. Nanomaterials Physicochemistry Department, Faculty of Chemical Technology and Engineering, West Pomeranian University of Technology, 71-065 Szczecin, Poland; Karolina.Wenelska@zut.edu.pl (K.W.); emijowska@zut.edu.pl (E.M.)
6. Indian Institute of Science Education and Research, Pune 411 008, India; aakanksha.kapoor@students.iiserpune.ac.in (A.K.); ashna@iiserpune.ac.in (A.B.)
7. Centre for Advanced Materials (CAM), Heidelberg University, INF 225, 69120 Heidelberg, Germany
* Correspondence: klingeler@kip.uni-heidelberg.de

Received: 3 February 2020; Accepted: 23 February 2020; Published: 27 February 2020

Abstract: Downsizing well-established materials to the nanoscale is a key route to novel functionalities, in particular if different functionalities are merged in hybrid nanomaterials. Hybrid carbon-based hierarchical nanostructures are particularly promising for electrochemical energy storage since they combine benefits of nanosize effects, enhanced electrical conductivity and integrity of bulk materials. We show that endohedral multiwalled carbon nanotubes (CNT) encapsulating high-capacity (here: conversion and alloying) electrode materials have a high potential for use in anode materials for lithium-ion batteries (LIB). There are two essential characteristics of filled CNT relevant for application in electrochemical energy storage: (1) rigid hollow cavities of the CNT provide upper limits for nanoparticles in their inner cavities which are both separated from the fillings of other CNT and protected against degradation. In particular, the CNT shells resist strong volume changes of encapsulates in response to electrochemical cycling, which in conventional conversion and alloying materials hinders application in energy storage devices. (2) Carbon mantles ensure electrical contact to the active material as they are unaffected by potential cracks of the encapsulate and form a stable conductive network in the electrode compound. Our studies confirm that encapsulates are electrochemically active and can achieve full theoretical reversible capacity. The results imply that encapsulating nanostructures inside CNT can provide a route to new high-performance nanocomposite anode materials for LIB.

Keywords: filled carbon nanotubes; lithium-ion batteries; hybrid nanomaterials; anode material

1. Introduction

Lithium-ion batteries (LIB) offer high gravimetric and volumetric energy densities which renders them particularly suitable for mobile applications. In order to optimize their performance, in particular with larger energy density, there is a continuous search for novel electrode materials. Electrode materials based on conversion and alloying mechanisms promise extremely enhanced electrochemical capacities in lithium-ion batteries as compared to conventional materials [1–3]. However, severe fading of the electrochemical capacity due to fractionation, resulting from pronounced volume changes upon electrochemical cycling, is one of the major drawbacks with respect to application. In addition to volume changes associated with the conversion reaction, low electric conductivity of many conversion materials seriously hinders their applicability in secondary batteries [4]. Nanosizing promises enhanced capability to accommodate strain induced by electrochemical cycling and may reduce kinetic limitations of the macroscopic counterparts of electrode materials [5–7] since downsizing particles yields shorter diffusion lengths and hence enhances rate performances of electrode materials. However, low density limiting volumetric energy densities of actual electrodes as well as high surface areas are relevant issues to be considered in nanomaterials as well. High reactivity associated with high surface area typically promotes irreversible processes and associated electrolyte consumption. In this respect, due to carbon's restricted voltage regime of electrochemical activity, carbon (nano) coating is a valuable tool to protect active nanomaterials, thereby avoiding enhanced electrolyte degradation and associated (and potentially dangerous) gas production [8]. Downscaling materials towards carbon-shielded hybrid nanomaterials hence offers a route to obtain electrode materials for LIB with enhanced performance.

Rational design of electrode materials has to tackle the abovementioned issues of low electronic conductivity limiting many promising electrode materials as well as of large volume changes during electrochemical cycling, with the latter particularly causing electrode structure and particles distortions and hence strong performance fading. Hierarchical nanocomposite carbon/active material structures offer an effective way to solve these issues as such materials exploit size effects of the nanoscaled building blocks [9–14]. Mechanical strain arising from volume changes is additionally buffered by the hierarchical structures. In this way, such materials optimally maintain the integrity of the bulk material while offering improved electrical conductivity owing to a carbon-based backbone structure [15–28]. Moreover, a strong backbone structure improves the stability of the composite with respect to mechanical strain arising from volume changes during electrochemical cycling.

We report CNT-based composite nanomaterials with enhanced electrochemical performance realized by filling material into CNT (for a schematics see Figure 1) which is electrochemically active when nanoscaled [30]. CNT display excellent conductivity as well as mechanical and chemical stability which renders them an excellent carbon source in hybrid nanomaterials [31]. However, in conventional approaches using exohedrally functionalized CNT, synthesis of uniformly sized and shape-controlled nanoparticles is challenging. In addition, while the interconnected network of carbon nanotubes provides an electrically conducting backbone structure, decorated nanoparticles onto the outer CNT-walls tend to lose electrical contact upon cycling-induced disintegration and particular methods have to be developed to improve connection to CNT [32–35]. Our results demonstrate successful synthesis of hybrid nanomaterial of CNT filled with Mn_3O_4, $CoFe_2O_4$, Fe_xO_y, Sn, and $CoSn$ and show the electrochemical activity of encapsulated materials. Encapsulates are either conversion or alloying electrode materials which perform the following general reactions upon electrochemical cycling, respectively [2,36,37]:

$$\text{Conversion: } M_aO_b + 2b\text{Li}^+ + 2b\text{e}^- \leftrightarrow aM^0 + b\text{Li}_2O \tag{1}$$

$$\text{Alloying: } M + x\,\text{Li}^+ + x\,\text{e}^- \leftrightarrow \text{Li}_xM \ (x < 4.25) \tag{2}$$

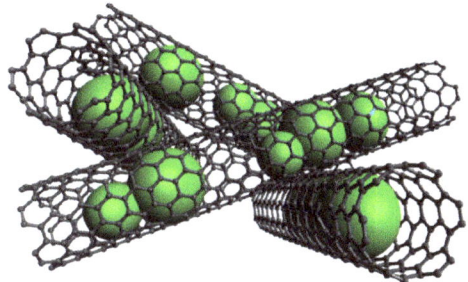

Figure 1. Schematics of nanocomposite material formed by interconnected carbon nanotubes (CNT) filled with high-capacity electrode materials. Essential characteristics are (1) size-controlled nanoparticles in the inner cavities of CNT which are separated from encapsulates in other CNT, (2) electrical contact of the incorporated material to a stable conductive network of CNT, (3) limitation of direct electrolyte/active material contact yielding and hence improved chemical stability. Created with Avogadro [29].

In this work, we demonstrate that in the case of conversion materials filled inside CNT, the encapsulated material completely participates in electrochemical cycling, i.e., the theoretical capacity is fully accessible. The backbone network of CNT is indeed unaffected by cracks of encapsulate which usually inhibit long-term stability. Our data hence imply that endohedrally functionalized CNT offer a promising route to new nanohybrid anode materials for LIB.

2. Synthesis and Characterization of Filled CNT

We report studies on hybrid nanomaterial of multiwalled carbon nanotubes (CNT) filled with Mn_3O_4, $CoFe_2O_4$, Fe_xO_y, Sn, and CoSn which have been fabricated by a variety of methods. Mostly, CNT of type PR-24-XT-HHT (Pyrograf Products, Inc., Cedarville OH, USA) have been used as templates. For introducing materials into the inner cavity of the CNT, mainly extensions of solution-based approaches reported in [38–43] have been applied [44,45]. This is illustrated by the example of Mn_3O_4@CNT which has been obtained by filling CNT with a manganese salt solution and a subsequent reducing step yielding homogeneously MnO-filled CNT (MnO@CNT) [4]. Subsequent heat treatment of MnO@CNT yields the complete conversion into Mn_3O_4@CNT, as confirmed by the XRD pattern in Figure 2. In case of filling with Co-Fe spinels, nitrate solutions of $Fe(NO_3)_3 \cdot 9H_2O$ (grade: ACS 99.0–100.2%) and $Co(NO_3)_2 \cdot 6H_2O$ (grade: ACS 98.0–102.0% metal basis) were used in stoichiometric ratios with respect to the metal ions (i.e., Fe:Co = 2:1). After adding CNT and treating the mixture in an ultrasonic bath with appropriate washing steps, the solid residue was dried and afterwards calcinated under argon flow atmosphere (100 sccm) at a temperature of 500 °C for 4 h to convert the nitrates into the corresponding cobalt ferrite. This is confirmed by XRD data in Figure 2 which indicate the presence of $CoFe_2O_4$. Pronounced peak broadening indicates the presence of nano-sized $CoFe_2O_4$ crystallites, with an estimated grain size of 20(5) nm by means of the Scherrer equation applied to the Bragg peak at 41.5°.

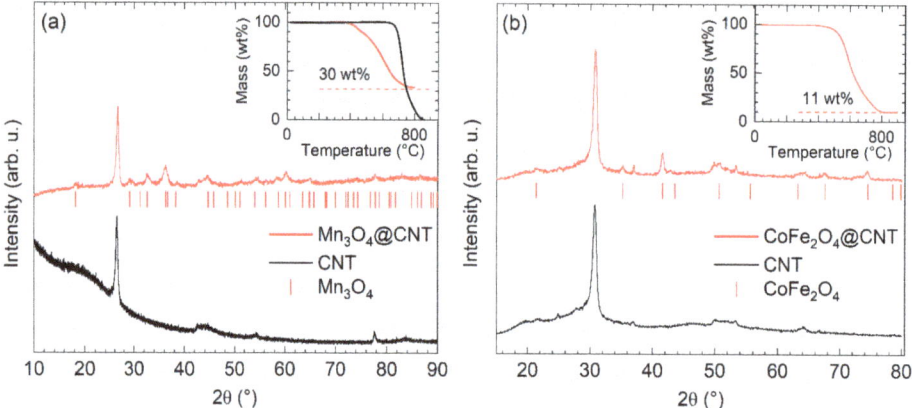

Figure 2. Left (**a**): PXRD patterns of Mn_3O_4@CNT and pure CNT. Vertical lines show the Bragg positions of Mn_3O_4 (space group $I4_1/amd$) [46]. Inset: Thermogravimetric analysis (TGA) data of Mn_3O_4@CNT and pure CNT. Right (**b**): PXRD patterns of $CoFe_2O_4$@CNT and of pristine CNT. Vertical ticks label Bragg positions of bulk $CoFe_2O_4$ (space group $Fd\bar{3}m$) [47]. Inset: TGA of $CoFe_2O_4$@CNT.

XRD patterns show relatively broad Bragg reflections which indicate small primary particle size of the noncarbon materials of the composite as expected for nanoparticles fitting inside the interior of CNT. This is confirmed by exemplary SEM and TEM studies presented in Figure 3. The images clearly show that the metal oxide nanoparticles are rather spherical and are located inside the CNT. Note the exception of possible nanowire formation in the case of metal-filled Sn@CNT as discussed in Section 3.4 (see Figure 15). The filling rate of Mn_3O_4@CNT is about 30(1) wt% and that of $CoFe_2O_4$@CNT (see the inset of Figure 2) is about 11(1) wt% as determined by thermogravimetric measurements (TGA).

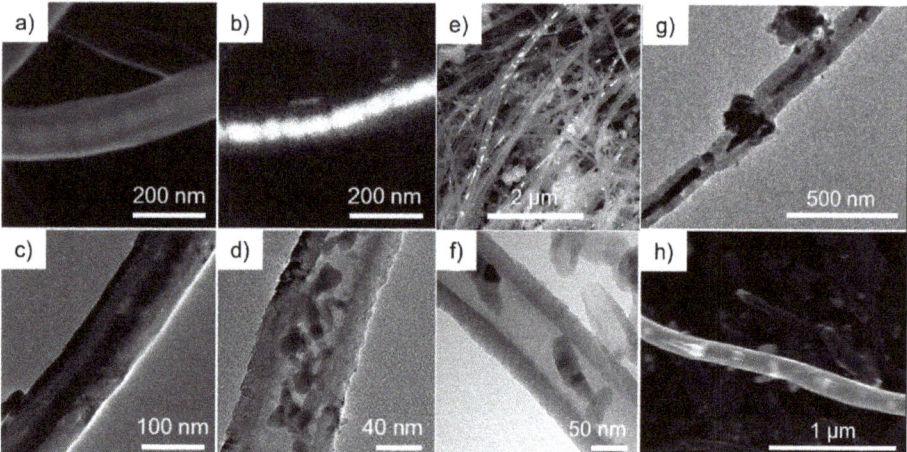

Figure 3. (**a**) SEM image of an individual Mn_3O_4@CNT (SE mode); (**b**) corresponding BSE mode image; (**c**,**d**) TEM images of different individual Mn_3O_4@CNT. Taken from [45]. (**e**) Overview SEM image of $CoFe_2O_4$@CNT (BSE mode); (**f**) TEM image of an individual $CoFe_2O_4$@CNT. (**g**) TEM image of an individual Fe_xO_y@CNT [48]. (**h**) SEM image of CoSn@CNT [49].

Electron microscopy confirms that the filling materials are located mainly inside the CNT. Exemplary SEM and TEM images are shown in Figure 3 (see also Figure 15 for Sn-filled CNT).

In Mn$_3$O$_4$@CNT, the encapsulated particles are rather spherical with the average diameter of 15 ± 7 nm obtained by TEM analysis. Note, that this is smaller than the size-limiting inner diameter of the utilized CNT (~35 nm). The SEM overview image (Figure 3e) on CoFe$_2$O$_4$@CNT also confirms that the filling material is distributed along the inner cavity of the hollow CNT. TEM indicates spherical encapsulates as well as short rods inside CNT (Figure 3e,f). Fe$_x$O$_y$@CNT (synthesis reported in [48]) appears to be mainly filled with α-Fe$_2$O$_3$ but also exhibits Fe$_3$O$_4$ as shown, e.g., by associated features in the magnetic susceptibility (see Section 3.3). Figure 3g also shows the presence of Fe$_x$O$_y$ nanoparticles outside CNT. In addition to separated spherical nanoparticles, encapsulates in CoSn@CNT and Sn@CNT form also nanowires up to 1 µm length (see Figure 3h and Figure 15). In either case, the encapsulates fill the complete inner diameter of the CNT, which is about 50 nm [44]. In summary, the results show that our synthesis approaches result in CNT filled with nanoparticles whose diameters are limited by the inner diameter of the CNT.

3. Electrochemical Studies

3.1. Mn$_3$O$_4$@CNT

Cyclic voltammetry studies on Mn$_3$O$_4$@CNT [30,45] and on pristine CNT, performed in the voltage range of 0.01–3.0 V vs. Li$^{0/+}$ and recorded at a scan rate of 0.1 mV s^{-1}, confirm electrochemical activity of encapsulates (Figure 4). During the initial cycle, starting with the cathodic scan, five distinct reduction peaks (R1–R5) and three oxidation peaks (O1–O3) are observed. The redox pair R1/O1 around 0.1 V and the irreversible reduction peak R3 at 0.7 V can be attributed to processes related to multiwalled CNT (Figure 4a). The irreversible reaction peak R3 signals formation of the solid electrolyte interphase (SEI) expected for carbon-based (here: CNT) systems [50]. The pronounced redox pair R1/O1 demonstrates that the bare CNT subsystem in the hybrid material is electrochemically active as it signals (de)lithiation of Li$^+$ ions between the layers of CNT [51,52]. Slight splitting of oxidation peak O1 indicates a staging phenomenon reported for graphite electrodes [37], and very similar behavior upon cycling is found in bare CNT [45]. All other features observed in Figure 4b are ascribed to the electrochemical reaction mechanism which has been reported for Mn$_3$O$_4$ as follows [53,54] (for further details see [45]):

(A) $Mn_3(\frac{1}{3} \cdot II, \frac{2}{3} \cdot III)O_4 + Li^+ + e^- \rightarrow LiMn_3(\frac{2}{3} \cdot II, \frac{1}{3} \cdot III)O_4$
(B) $LiMn_3O_4 + Li^+ + e^- \rightarrow Li_2O + 3 \cdot Mn(II)O$
(C) $Mn(II)O + 2 \cdot Li^+ + 2 \cdot e^- \leftrightarrow Li_2O + Mn(0)$

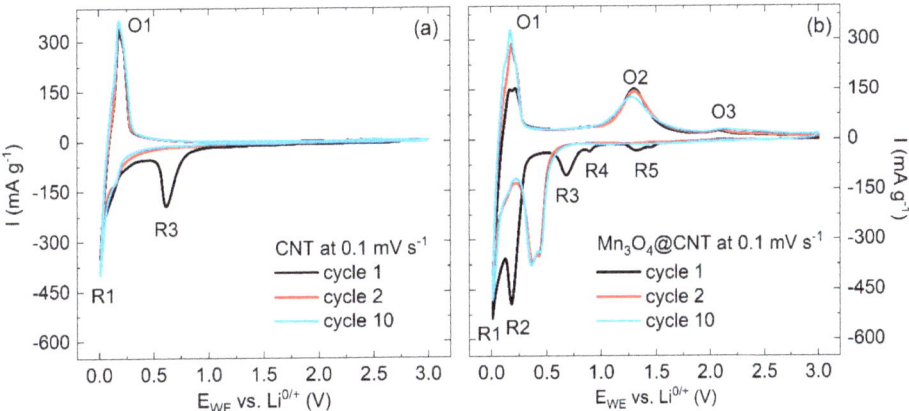

Figure 4. Cyclic voltammograms of (**a**) pristine CNT and (**b**) Mn$_3$O$_4$@CNT at 0.1 mV s^{-1} [30].

The cyclic voltammograms (CVs) confirm electrochemical activity of encapsulated Mn_3O_4. Absence of significant changes between cycles 2 and 10 indicate good cycling stability which will be investigated in more detail below. Since the materials associated with the mechanism detailed in Equations (A) to (C) exhibit strong differences in magnetic properties, magnetic studies are suitable to follow the redox reaction. In particular, there are strong changes of magnetic properties upon electrochemical cycling from ferrimagnetic Mn_3O_4 to antiferromagnetic MnO (Figure 5; for further magnetization data see [45]). Pristine Mn_3O_4@CNT shows ferrimagnetic order below $T_C = 42$ K as indicated by the magnetization data. In contrast, materials extracted after step (B) of the abovementioned redox reactions, i.e., after galvanostatic reduction at 5 mA g^{-1} down to 0.5 V and passing the reduction peaks R5, R4, and R3 labelled in Figure 4b, displays nearly no traces of ferrimagnetic material. Quantitatively, the magnetization data indicate about 1% remainder of ferrimagnetic Mn_3O_4 after the first half cycle. Meanwhile, antiferromagnetic order is found below a temperature of ~120 K, which is expected for MnO [55] and is in agreement with Equation (B). Hence, our magnetometry data confirm electrochemical reactions as postulated in Equations (A)–(C) by tracking down individual magnetic species.

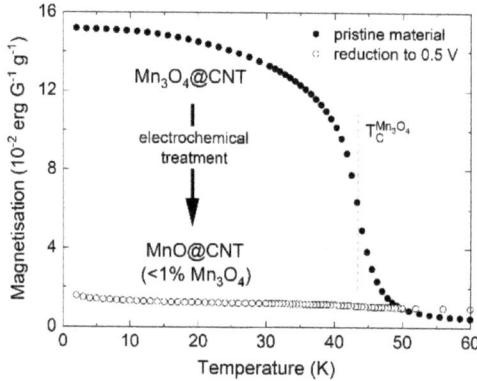

Figure 5. Magnetization of pristine and electrochemically cycled Mn_3O_4@CNT measured at B = 0.1 T (FC). The vertical line indicates the ferrimagnetic ordering temperature in Mn_3O_4.

Charge and discharge studies at specific current rates (Figure 6) display plateau-like regions in the voltage profiles signaling the redox features discussed above by means of Figure 4. In the initial cycle performed at 50 mA g^{-1}, specific charge and discharge capacities of 677 and 455 mAh g^{-1}, respectively, are achieved. Increasing the charge/discharge current to 100 and 250 mA g^{-1}, respectively, does not significantly affect the shape of the curves but yields smaller discharge capacities, e.g., 331 mAh g^{-1} after 30 cycles. For higher currents, the plateaus corresponding to delithiation and lithiation of CNT vanish, while the conversion reaction (Equation (C)) is still visible in the data. The rate capability studies presented in Figure 6 display pronounced capacity losses when increasing charge/discharge currents. Specifically, maximum discharge capacities of 468, 439, 349, 245, and 148 mAh g^{-1} are reached at 50, 100, 250, 500 and 1000 mA g^{-1}, respectively.

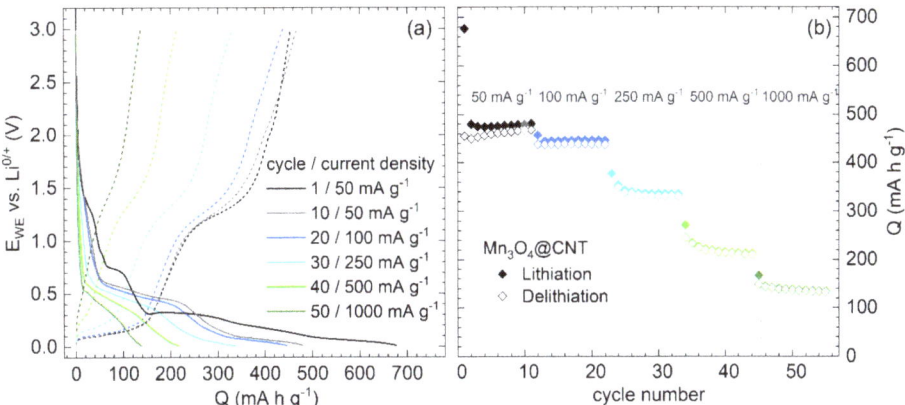

Figure 6. Rate capability studies of Mn_3O_4@CNT at 50, 100, 250, 500, and 1000 mA g^{-1}. (**a**) Potential profiles of specific lithiation (solid lines) and delithiation cycles (dashed lines). (**b**) Specific charge/discharge capacities upon cycling [30].

In order to assess the electrochemical performance of the composite with particular emphasis on the encapsulate, evolution of capacities at 100 mA g^{-1} (galvanostatic cycling with potential limitation) upon cycling of Mn_3O_4@CNT and pristine CNT is shown in Figure 7. While the initial half cycle is strongly affected by irreversible processes associated with solid electrolyte interface (SEI) formation, the Mn_3O_4@CNT nanocomposite exhibits increasing capacities for approximately 15 cycles in contrast to decreasing values of pristine CNT. The nanocomposite reveals a maximum discharge capacity of 463 mA h g^{-1} in cycle 18, of which 93% is maintained after 50 cycles (429 mA h g^{-1}). Thus, incorporation of Mn_3O_4 into CNT leads to more than 40% enhanced specific capacities on average as compared to unfilled CNT. The data, i.e., on filled and unfilled CNT, enable calculating the specific capacity of incorporated Mn_3O_4 (29.5 wt%). The encapsulate's initial capacity of about 700 mAh g^{-1} increases significantly to 829 and 820 mAh g^{-1} (cycle 18) and declines thereafter, with capacity retention of around 90% after 50 cycles. The Mn_3O_4 capacity even exceeds the theoretical expectations of the conversion reaction (C) from cycle 6 on (dashed line in Figure 7). This might be associated with a capacity contribution due to oxidative feature O3 (Figure 4b), which supposedly indicates the back-formation of Mn_3O_4 and corresponding reduction processes [56,57]. Note, however, the error bars of 5% due to mass determination of encapsulate and subtraction of data on pristine CNT. Initial capacity increase was also observed in previous studies on Mn_3O_4/CNT composites [58,59].

Our analysis shows that full conversion between MnO and metallic Mn can be achieved reversibly and the maximum of the contributed capacity by the Mn_3O_4 encapsulate is accessible (Figure 7). In particular, the nanoparticles inside CNT are completely involved in the electrochemical processes. This finding is supported by the fact that the active material inside CNT experiences distinct structural changes, as evidenced by TEM studies (Figure 8). Figure 8b,c presents materials after 13 galvanostatic cycles, at 100 mA g^{-1}, taken after delithiation and lithiation. No clear differences are observed between the lithiated and the subsequently delithiated material. In both cycled materials, the encapsulate which initially exhibits well-defined, rather spherical nanoparticles has developed extended patches. The TEM image also shows lower contrast of the encapsulate to the CNT environment which is indicative of lower density of the encapsulate. Equations (A) and (B) indeed suggest rather larger volume expansion of Mn_3O_4 during initial lithiation and concomitant agglomeration as well as amorphization of the filling which is in agreement with the TEM results. Notably, despite the strong changes of encapsulate, CNT mantles still display the characteristic graphitic layers of multiwalled carbon nanotubes (see Figure 8d). Hence, electrochemical cycling does not severely damage the structure of the CNT. Furthermore, an amorphous layer of ~5 nm thickness can be observed on top of the graphitic CNT layers, which can be attributed to the SEI. The TEM analysis hence shows that the CNT indeed offer a stable environment for the manganese oxides which is able to accommodate the strain due to volume expansion during electrochemical cycling and guarantees a consistent electrical contact to the active material.

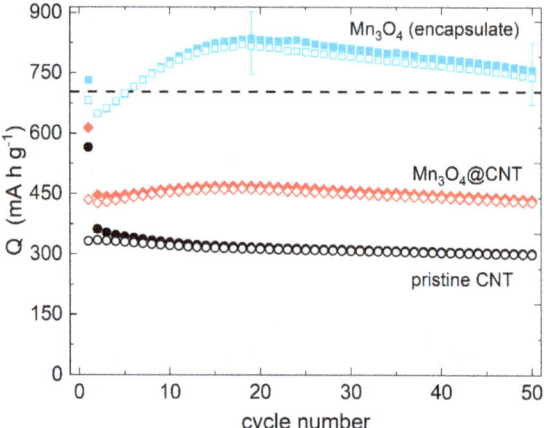

Figure 7. Specific charge/discharge capacities at 100 mA g^{-1} of pristine CNT, Mn$_3$O$_4$@CNT, and calculated capacity of the encapsulate. The dashed line shows the theoretical capacity of the reversible conversion reaction (C) [30].

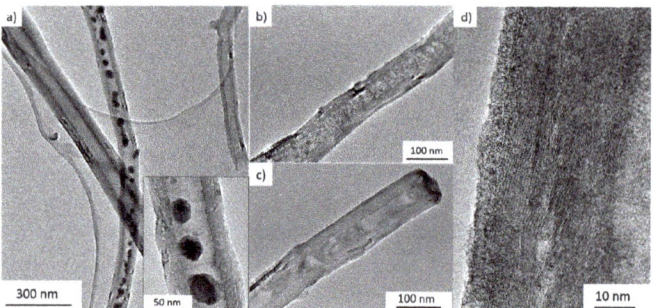

Figure 8. TEM images of (**a**) uncycled, (**b**) galvanostatically lithiated, and (**c**) delithiated Mn$_3$O$_4$@CNT. (**d**) High-resolution TEM image of a CNT shell of delithiated material after 13 cycles. Taken from [45].

3.2. CoFe$_2$O$_4$@CNT

As shown in Figure 3f,g, nanosized particles of cobalt ferrite CoFe$_2$O$_4$ are incorporated into CNT by a similar procedure as applied in the case of Mn$_3$O$_4$@CNT. The mass content of CoFe$_2$O$_4$ in the composite materials however amounts to only 11 wt%, leading to smaller effects of the encapsulate. In order to evaluate the benefits of CNT shells, the electrochemical performance of the nanocomposite CoFe$_2$O$_4$@CNT is compared to that of bare CoFe$_2$O$_4$ nanoparticles (Figure 9). In general, electrochemical lithium storage of up to 8 Li^1/f.u. in CoFe$_2$O$_4$ follows a conversion mechanism (Equation (D)), which may be preceded by initial intercalation of Li$^+$ ions into the original ferrite structure [60]:

(D) $CoFe_2O_4 + 8\ Li^+ + 8\ e^- \rightarrow Co + 2\ Fe + 4\ Li_2O$
(E) $Co + 2\ Fe + 4\ Li_2O \leftrightarrow CoO + Fe_2O_3 + 8\ Li^+ + 8\ e^-$

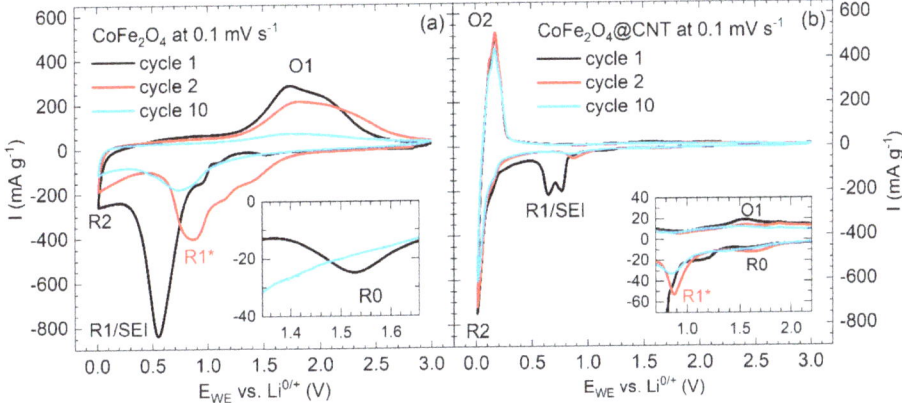

Figure 9. Cyclic voltammograms of (**a**) pristine $CoFe_2O_4$ and (**b**) $CoFe_2O_4$@CNT, at 0.1 mV s^{-1} [30].

Both processes show up as redox features in the CVs in Figure 9 which for $CoFe_2O_4$@CNT also show features present in pristine CNT (Figure 4a) [45,51,61,62].

In bare $CoFe_2O_4$ nanoparticles (Figure 9a), the initial half cycle reduction peaks indicate, at 1.5 V, initial intercalation into the spinel structure (R0), and at 1.1 and 0.55 V indicate R1/SEI formation. In addition, there is a shoulder at 0.95 V and a peak at 0.01 V (R2). In all subsequent reductive half cycles, the most pronounced reduction peak occurs at 0.85 V (R1*). Expectedly, R0 vanishes after the first cycle. The oxidative scans display a broad oxidation double peak between 1.5 V and 2.5 V with a maximum intensity around 1.65 V (O1). R1 most likely indicates both conversion of the spinel to Co and Fe [60,63] and SEI formation [64], while R2 signals intercalation of Li$^+$ ions into added carbon black [64,65]. Upon further cycling, Co and Fe oxidize to CoO and Fe_2O_3, respectively (O1), followed by the corresponding conversion processes at R1* (Equation (D)) [36,63,66–69].

CVs on $CoFe_2O_4$@CNT in Figure 9b show features associated with $CoFe_2O_4$ superimposed by redox peaks related to CNT. In the initial cycle, features attributed to $CoFe_2O_4$ appear at 1.6 (R0), 1.2, and 0.7 V (SEI) with a shoulder at 0.8 V (R1). Upon further cycling, they are shifted to 1.6 (R0) and 0.9 V (R1*). Reversible oxidation peaks appear at similar voltages as compared to bare $CoFe_2O_4$ nanoparticles, i.e., between 1.5 and 2.0 V with a maximum at 1.55 V (O1). The results imply smaller overpotentials in $CoFe_2O_4$@CNT as compared to the bare $CoFe_2O_4$ nanoparticles, indicating improved energy efficiency. Furthermore, cycling stability is superior, yielding noticeable redox activity of the $CoFe_2O_4$ encapsulate in the 10th cycle. Both improvements can be attributed to benefits of the $CoFe_2O_4$@CNT composite material, i.e., to enhanced overall conductivity and better structural integrity.

These conclusions are corroborated by galvanostatic cycling with potential limitation (GCPL) data (Figure 10). Firstly, higher capacities of $CoFe_2O_4$@CNT as compared to pristine CNT imply electrochemical activity of encapsulates for 60 cycles under study. In addition to irreversible effects associated with SEI formation, there are capacity losses, in particular in initial cycles, so that the electrode demonstrates only 97% of Coulombic efficiency after 15 cycles. Capacity retention of $CoFe_2O_4$@CNT amounts to a fair value of 76% after 60 cycles (243 mAh g^{-1}). Analogously to Section 3.1, the specific contribution of $CoFe_2O_4$ is evaluated by subtracting the measured capacities of pristine CNT, weighted with the mass ratio of 89:11 (CNT:$CoFe_2O_4$). The analysis shows (Figure 10b) that both for pristine and CNT-encapsulated $CoFe_2O_4$ there are pronounced capacity losses upon cycling while the initial capacities exceed the theoretical maximum value of 914 mAh g^{-1} due to SEI formation. CNT-encapsulated active material clearly outperforms bare $CoFe_2O_4$ nanoparticles. To be specific, after 20 cycles, 475 mAh g^{-1} (71%) is retained in $CoFe_2O_4$@CNT while the bare particles show 190 mAh g^{-1} (22%). This result again demonstrates that embedding nanosized $CoFe_2O_4$ inside CNT partly compensates for the typical capacity fading associated with the conversion reactions upon electrochemical delithiation or lithiation known for spinel materials.

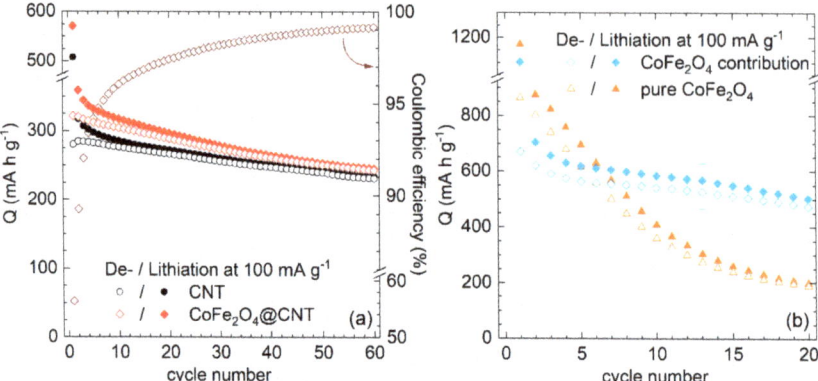

Figure 10. (a) Specific charge/discharge capacities, at 100 mA g^{-1}, of pristine CNT and CoFe$_2$O$_4$@CNT as well as the Coulombic efficiencies of the latter. (b) Capacity contribution of the encapsulated CoFe$_2$O$_4$ in comparison to pristine CoFe$_2$O$_4$ [29].

While encapsulated CoFe$_2$O$_4$@CNT demonstrates electrochemical activity, it is illustrative to compare the results with alternative carbon/CoFe$_2$O$_4$ hybrid nanomaterials. Direct comparison is often hindered by the fact that the carbon-related capacity is not always subtracted as done here. For many carbon/CoFe$_2$O$_4$ hybrid materials, much higher values than maximum theoretical capacity of CoFe$_2$O$_4$ are reported. A value of 1046 mAh g^{-1} is reported for mesoporous CoFe$_2$O$_4$ nanospheres cross-linked by carbon nanotubes [70]. Porous carbon nanotubes decorated with nanosized cobalt ferrite show 1077 mAh g^{-1}, after 100 cycles [69]. More than 700 mAh g^{-1} of total capacity of the composite was obtained when CoFe$_2$O$_4$ is encapsulated into carbon nanofibers with 36% carbon content [71]. A list of recently achieved record values may be found in [72]. We note that excessive capacity beyond the theory values in transition metal oxide/carbon nanomaterials have been associated, e.g., to decomposition of electrolyte and formation of a polymer/gel-like film on the nanoparticles [73]. Another hypothesis refers to interface charging effects by lithium accommodation at the metal/Li$_2$O interface [74]. Our data indeed suggest that surface effects might be relevant as CNT-encapsulation of active material evidently suppresses this phenomenon.

3.3. Fe$_x$O$_y$@CNT and CNT@Co$_3$O$_4$

Fe$_x$O$_y$@CNT has been synthesized as described in [48]. XRD and magnetic characterization studies [30] imply the presence of several iron oxides (i.e., of α-Fe$_2$O$_3$ as well as of γ-Fe$_2$O$_3$ or/and Fe$_3$O$_4$) in the materials. While the main phase appears as α-Fe$_2$O$_3$, magnetic studies show both the Morin and Verwey transitions which enable to unambiguously identify α-Fe$_2$O$_3$ and Fe$_3$O$_4$, respectively. Note, that the presence of antiferromagnetic γ-Fe$_2$O$_3$ can neither be confirmed nor excluded by our magnetic studies. Analyzing the magnetization data indicates the presence of ferromagnetic iron oxide (i.e., γ-Fe$_2$O$_3$ and/or Fe$_3$O$_4$) of about 30(8) wt%.

The CVs shown in Figure 11a display two reductions (R1, R2) and two oxidations (O1, O2) which are observed in all cycles. We attribute R1/O1 to electrochemical activity of CNT. Except for typical initial irreversible effects at R2/SEI, all features are well explained by electrochemical processes known in iron oxides. Mechanisms in α-Fe$_2$O$_3$ as identified by Larcher et al. [75,76] involve Li-intercalation in nanoparticles, followed by conversion to metallic Fe and Li$_2$O via intermediately formed cubic Li$_2$Fe$_2$O$_3$. This process is partly reversible as it includes formation of FeO [77] and γ-Fe$_2$O$_3$ [78,79]. For Fe$_3$O$_4$, after initial intercalation, Li$_2$Fe$_3$O$_4$ is formed which is subsequently reduced to Fe and

Li$_2$O [80,81]. In all iron oxides present in Fe$_x$O$_y$@CNT, including γ-Fe$_2$O$_3$, electrochemical processes display similar features which are not well distinguishable [82,83]. The inset of Figure 11a presents a weak reduction peak R3 which we attribute to abovementioned Li-intercalation into iron oxides. Note, that the second peak in the inset is due to an intrinsic cell setup effect. Conversion reactions appear at around 0.6 V and are signaled by feature R2. The shoulder at 0.8 V indicates the successive nature of the lithiation processes. Upon cycling, R2 shifts to 0.9–1.2 V, thereby indicating significant structural changes due to the initial conversion process. The large width of O2 might indicate several oxidation processes upon delithiation. The evolution of the oxidation features upon cycling implies severe fading effects.

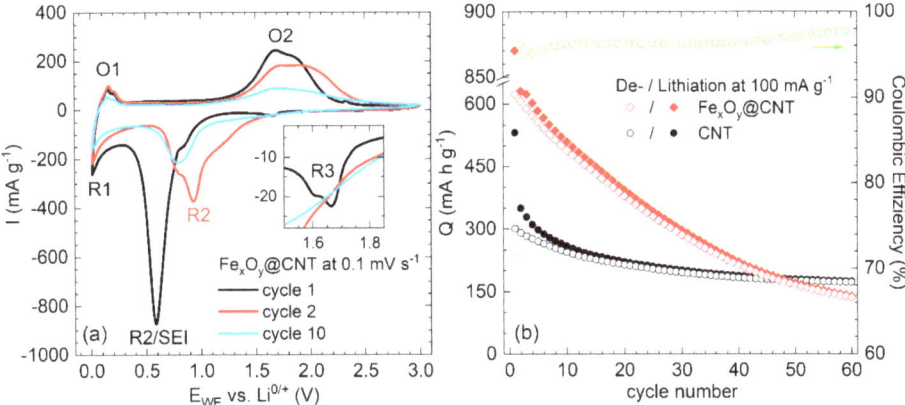

Figure 11. (a) Cyclic voltammogram of Fe$_x$O$_y$@CNT at 0.1 mV s^{-1}. (b) Specific charge/discharge capacities, at 100 mA g^{-1}, of pristine CNT (Pyrograf Products, type PR-24-XT-HHT) and Fe$_x$O$_y$@CNT, as well as the Coulombic efficiencies of the latter [30].

This is confirmed by the data in Figure 11b which presents specific charge/discharge capacities of Fe$_x$O$_y$@CNT obtained at 100 mA g^{-1}. Respective data on bare CNT (Pyrograf Products, type PR-24-XT-HHT) are shown for comparison. The initial capacities of the composite amount to 870 and 624 mAh g^{-1}, which reflects initial irreversible processes. There is a strong decay in capacity which yields only 78% (489 mAh g^{-1}) in cycle 10 and 26% (165 mAh g^{-1}) in cycle 50 of the initial discharge capacity. The results clearly show that envisaged improvement of cycling stability due to encapsulation into CNT is not achieved. Presumably, iron oxide content outside CNT is rather large so that a significant part of functionalization is exohedral. In such case, we assume that volume changes upon cycling leads to detachment of these particles from the CNT network which results in diminished electrochemical activity. In contrast, [84] reports α-Fe$_2$O$_3$-filled CNT which show 90% capacity retention in cycle 50.

Inferior stability of exohedrally functionalized CNT upon electrochemical cycling is further confirmed, e.g., for CNT decorated by mesoporous cobalt oxide (CNT@Co$_3$O$_4$). The material was synthesized as reported in [34]. The composite exhibits 41 wt% of mesoporous Co$_3$O$_4$ spheres with mean diameters between 100 and 250 nm decorated to the CNT network. The electrochemical behavior of CNT@Co$_3$O$_4$ (Figure 12a) during the initial cycle shows SEI formation and the initial reduction process of Co$_3$O$_4$ to metallic cobalt and formation of amorphous Li$_2$O during the initial cycle. Double peaks appearing in cycle 2 correspond to a multistep redox reaction caused by the Co^{2+}/Co^0 and Co^{3+}/Co^{2+} couples [85,86]. The integrated specific capacities calculated from the CVs (Figure 12b) display significant capacity fading upon continued cycling. For comparison, a blend of separately fabricated CNT and Co$_3$O$_4$ nanoparticles were mechanically mixed postsynthesis in the same ratio of 59% CNT and 41% Co$_3$O$_4$ which, according to TGA, is realized in the decorated CNT@Co$_3$O$_4$ nanocomposite. The blend's CV shows similar peak positions as found in CNT@Co$_3$O$_4$, and a similarly high reductive capacity is measured for the postsynthesis blend in the first cycle. However, the associated reversible capacity is much lower as compared to the CNT@Co$_3$O$_4$ hybrid nanomaterial and the irreversible loss between charge and discharge capacity is higher. After a few cycles, both the blend and CNT@Co$_3$O$_4$ show similarly low performance, which indicates that the benefit of attaching mesoporous Co$_3$O$_4$ to the surface of CNT has completely faded, presumably due to detachment of the mesoporous Co$_3$O$_4$ nanospheres [34].

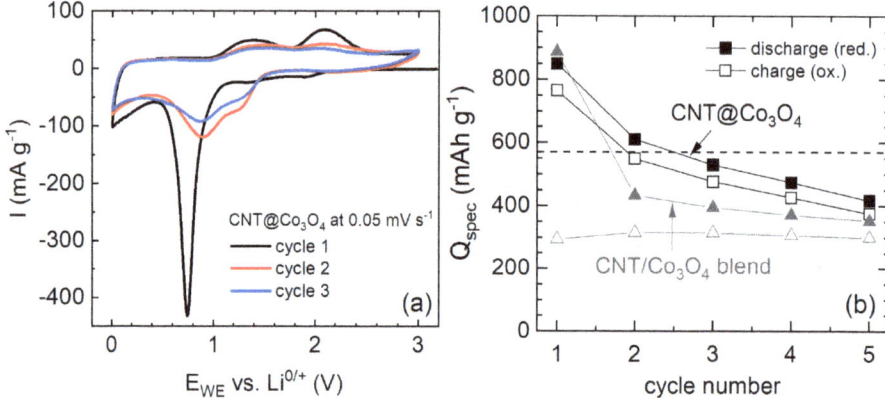

Figure 12. (a) CV curves of CNT@Co$_3$O$_4$, at 0.05 mV s^{-1} in the voltage range of 0.01–3.00 V. (b) Integrated charge and discharge capacities for five cycles as deduced from CV [32].

3.4. Sn@CNT and CoSn@CNT

The alloying process described by Equation (2) implies feasibility of (semi)metallic electrode materials for electrochemical energy storage. Using M = Ge, Sn, the alloy Li$_x$M is formed with x up to 4.25 Li$^+$/f.u. [87,88]. While Ge exhibits lower molecular weight and good Li$^+$-diffusivity, Sn is much cheaper and exhibits higher electrical conductivity [89]. For Sn, the most Li-rich alloy is Li$_{17}$Sn$_4$ (= Li$_{4.25}$Sn) which implies a theoretical capacity of 960 mAh g^{-1} [90]. Upon lithiation, several stable alloys such as LiSn and Li$_7$Sn$_2$ are formed, resulting in complex (de)alloying processes of several stages which are associated with large volume changes [91]. In CoSn@CNT, Co is electrochemically inactive and is supposed to buffer the volume changes as similarly done in a commercial Sn-Co-C composite by Sony [87,92,93].

Synthesis of Sn@CNT has been published in [44]. While the encapsulate in Sn@CNT is β-Sn with a filling ratio of 20 wt%, encapsulate in CoSn@CNT is a mixture of β-Sn, CoSn, and mainly CoSn$_2$ with in total 17 wt% of Sn and 5 wt% of Co. In addition to encapsulated separated spherical nanoparticles, encapsulates in Sn@CNT also form nanowires up to 1 μm in length. Both spheres and wires fill the complete inner diameter of the CNT, which is about 50 nm [44].

The CVs of Sn@CNT- and CoSn@CNT-based electrode materials shown in Figure 13 are similar to each other and confirm the multistage processes expected from reports on non-CNT materials. In both cases, in addition to the SEI formation, the peaks R1/O1 signal electrochemical activity of CNT. The reduction peak R3 at 0.6 V and the pair R2/O2 at 0.3 V as well as several features at 0.35–0.85 V, are all attributed to multi-stage (de)alloying processes.

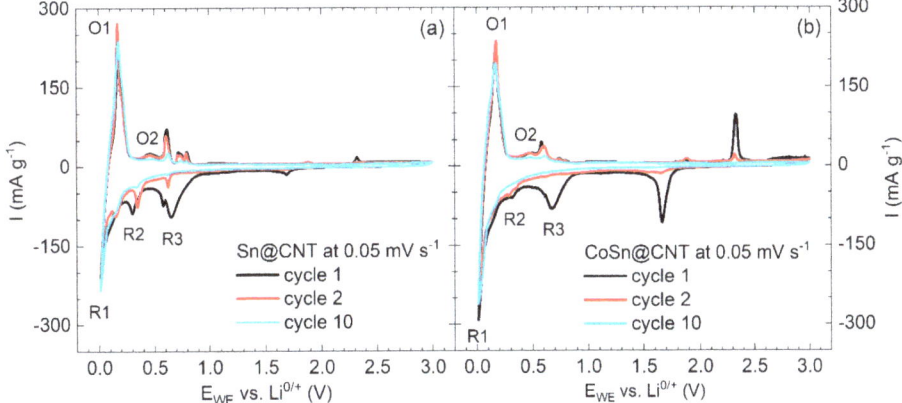

Figure 13. CVs of Sn@CNT (**a**) and CoSn@CNT (**b**) in the regime 0.01–3.0V vs. Li/Li+ at a scan rate 0.05 mV/s. Note that the oxidation peaks at 1.9 and 2.3 V and the reduction peak at 1.7 V appearing in the first two cycles are due to the experimental cell setup [30].

Galvanostatic cycling of Sn@CNT and CoSn@CNT as compared to pristine CNT (Pyrograf Products, Typ PR-24-XT-HHT) quantifies the contribution of encapsulates to the materials' capacities (Figure 14a). Sn@CNT displays clearly improved values. Quantitatively, the data imply an initial reversible capacity of 322 mAh g^{-1} in cycle 2, of which 281 mAh g^{-1}, i.e., 87%, is retained in cycle 50. In contrast, fading is much more severe in CoSn@CNT, which shows only 66% retained of the initial capacity 317 mAh g^{-1}, i.e., its performance in cycle 50 falls below that of pristine CNT. As will be discussed below, these data show that there is no positive (buffering) effect of alloyed Co. Rate capacities shown in Figure 14b at cycling rates 100–2000 mA g^{-1} illustrate the strong effect of fast cycling on both materials, thereby confirming limiting kinetics of the underlying electrochemical alloying processes.

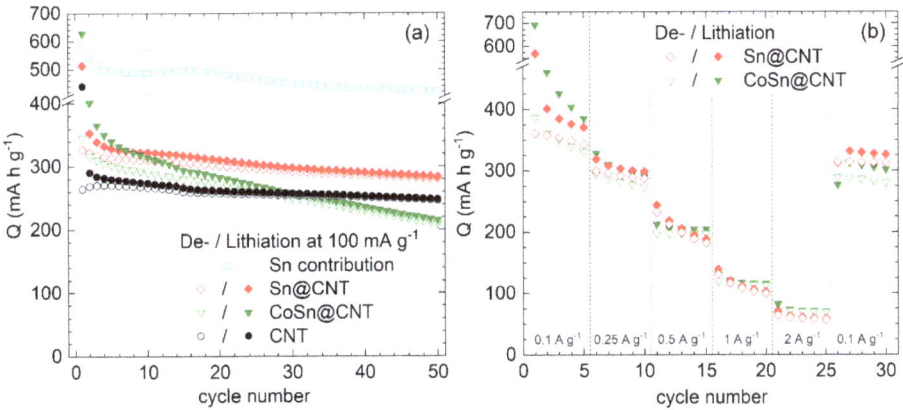

Figure 14. Specific capacities of Sn@CNT-, CoSn@CNT-, and pristine CNT-based electrodes. (**a**) Galvanostatic cycling with potential limitation (GCPL) at 100 mA g^{-1}. Blue data markers show the specific capacity of the Sn encapsulate after correcting the contribution of CNT. (**b**) GCPL at different rates of 0.1–2.0 A g^{-1} [30].

Figure 14a also presents the specific capacity of the Sn encapsulate which is derived by correcting the data by the effect of pristine CNT. Note error bars of Sn-capacity of up to 20% resulting in particular from errors in determining the filling ratio. In the first cycle, the reversible capacity amounts to 589 mAh g^{-1} which suggests deintercalation of x = 2.6 Li$^+$/f.u. Capacity fading is about 15% between cycle 5 (495 mAh g^{-1}) and cycle 50 (422 mAh g^{-1}). Even the initial capacities are much smaller than the theoretical one of 960 mAh g^{-1} that would be achieved for x = 4.25.

In-situ XRD studies on Li$_x$Sn$_y$ have shown that intermediate phases Li$_2$Sn$_5$ and LiSn are expected [94]. In agreement with these studies, the presence of (at least) two reduction peaks in the CVs of both materials (see Figure 13) suggests at least a two-stage process in the materials at hand. However, comparison to the literature does not allow to attribute these peaks to a specific process. This also holds for the observed (at least) four oxidation peaks which indicate step-wise delithiation of the Li$_x$Sn$_y$-alloy. For CoSn@CNT where the encapsulate mainly consists of CoSn$_2$, Mössbauer studies have shown the formation of Li$_x$Sn with x $\approx$ 3.5 in the first cycle [95]. Such a process is not visible in the CV (Figure 13b) but the respective feature might be masked by the SEI-peak. It is argued in [95] that, upon delithiation, Li$_{-3.5}$Sn forms an amorphous Li$_x$Co$_y$Sn$_2$-matrix which is crucial for the expected buffering associated with Co-alloying. We conclude that, in CoSn@CNT, this Li$_x$Co$_y$Sn$_2$-matrix is not realized but Co just deteriorates the electrochemical performance. This conclusion is supported by the fact that the CVs in Figure 13 display the same number of peaks at similar potentials in both Sn@CNT and CoSn@CNT, which indicates identical processes. We assume separation of Co and Sn instead of Li$_x$Co$_y$Sn$_2$-formation yielding electrochemically inactive regions.

The effect of galvanostatic cycling, at 50 mA g^{-1}, on Sn@CNT is demonstrated by TEM images in Figure 15. In the cycled materials, well separated homogenous encapsulates (Figure 15a,b) in the pristine material convert to rather completely but inhomogeneously filled CNT whose filling is indicated by different TEM contrast, i.e., different densities of encapsulate. These finding agrees with expected volume changes, in particular to large expansion upon lithiation, and phase separation of encapsulated material. One may speculate that the dark regions visible after cycling indicate electrochemically-inactive domains of Sn. The presence of inactive regions would be in agreement to the GCPL data (Figure 14) which show that only a maximum of 60% of the full Sn-capacity is achieved. Finally we note that previous studies on Sn-filled CNT have demonstrated better performance as compared to the material at hand. Wang et al. have reported Sn@CNT with filling ratios of 38 wt% and 87 wt% [96]. The former, i.e., less filled, CNT have demonstrated superior performance with capacities of 500 mAh g^{-1} for 80 cycles at 100 mA g^{-1}. The relevant parameter seems to be the size of Sn particles, which was 6–10 nm in [96]. Larger encapsulates filling the complete inner diameter of the CNT of about 50 nm as realized in the materials at hand seem to be detrimental and may cause electrochemically inactive regions. Addition of Co as a potential buffer does not improve the performance of such rather large nanoparticles inside CNT but even causes additional capacity fading.

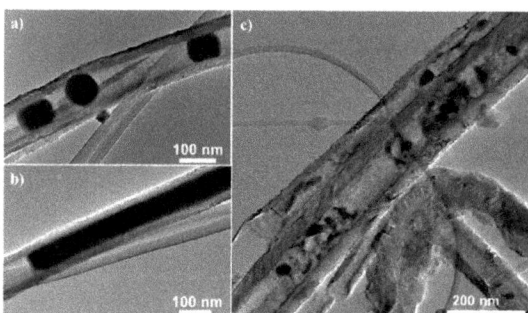

Figure 15. TEM images of pristine (**a**,**b**) and galvanostatically cycled (10 cycles) Sn@CNT (**c**).

4. Experimental Methods

4.1. Material Characterization

Materials were characterized by X-ray diffraction (XRD) with either Stadi P (Stoe, Darmstadt, Germany) using Cu Kα1 radiation (λ = 1.5406 Å) or X'Pert Pro MPD PW3040/60 (PANanalytical, Almelo, Niederlande) using Co Kα radiation (λ = 1.79278 Å). Thermogravimetric analysis (TGA) was carried out with SDT Q600 (TA Instruments, Waters Corporation, Milford, MA, USA). The morphology was investigated by means of scanning electron microscopy (SEM, Nova NanoSEM 200 (FEI company, Hilsboro, Oregon, USA)) and transmission electron microscopy (TEM, JEM-2010F (JEOL, Akishima, Japan), Tecnai (FEI company, Hilsboro, Oregon, USA)). A MPMS-XL5 (Quantum Design, San Diego, Californis, USA) superconducting quantum interference device (SQUID) magnetometer was used to perform magnetic measurements.

4.2. Electrochemical Measurements

Electrochemical properties were studied by cyclic voltammetry (CV) and galvanostatic cycling with potential limitation (GCPL) in Swagelok-type cells [97]. The measurements were performed on a VMP3 potentiostat (BioLogic) at a temperature of 25 °C. For the preparation of the working electrode, the active material was optionally mixed with carbon black (Super C65, Ymeris Graphite and Carbon, Bironico, Switzerland) and stirred in a solution of polyvinylidene fluoride (PVDF, Solvay, Brussels, Belgium) in N-methyl-2-pyrrolidone (NMP) for at least 12 h. After evaporating most of the NMP in a vacuum oven (80 °C, <10 mbar) the spreadable slurry was applied on copper mesh current collectors (Ø 10 mm). The as-prepared electrodes were dried at 80 °C in a vacuum oven (<10 mbar), mechanically pressed at 10 MPa, and afterwards dried again. The assembly of cells was done in a glovebox under argon atmosphere (O_2/H_2O < 5 ppm) using a lithium metal foil disk (Alfa Aesar, Haverhill, MA, USA) pressed on a nickel current collector as counter electrode. The electrodes were separated by two layers of glass microfibre (Whatman GF/D) soaked with 200 µL of a 1 M $LiPF_6$ salt solution in 1:1 ethylene carbonate and dimethyl carbonate (Merck ElectrolyteLP30). For post cycling studies, working electrodes were washed three times in dimethyl carbonate and afterwards dried under vacuum.

5. Conclusions

Endohedral functionalization of multiwalled carbon nanotubes by means of high-capacity electrode materials is studied with respect to application for electrochemical energy storage. Encapsulation indeed yields size-controlled nanoparticles inside CNT. The presented data imply that the filled materials are electrochemically active and can achieve full theoretical reversible capacity. While conversion and alloying processes yield cracks and amorphization of the encapsulate, the CNT mantles are found to be only very little affected by electrochemical cycling. The backbone network of CNT hence maintains its integrity and improved performance with respect to unshielded or exohedrally-attached nanomaterials. For appropriately tailored materials, CNT-based nanocomposites show smaller overpotentials and hence improved energy efficiency as well as improved cycling stability. The results imply that encapsulating nanostructures inside CNT provides a successful route to new high-performance nanocomposite anode materials for LIB.

Author Contributions: Conceptualization, E.T., A.O. and R.K.; methodology, E.T., A.O., S.H., A.B., E.M. and R.K.; data analysis, E.T., A.O. and R.K.; synthesis and material characterization, R.G., M.G., C.N. (Christian Nowka), M.S., M.H., S.W., S.H., A.K., A.B., K.W. and E.M.; electrochemical investigations, E.T., A.O., P.S., L.M., L.D., R.Z., F.W., L.S. and C.N. (Christoph Neef); writing—original draft preparation, E.T., A.O., and R.K.; writing—review, E.T., A.B., E.M., C.N., A.O, S.H., S.W. and R.K. All authors have read and agreed to the published version of the manuscript.

Funding: Work was partly supported by Deutsche Forschungsgemeinschaft DFG via KL 1824/12-1 and the CleanTech-Initiative of the Baden-Württemberg-Stiftung (Project CT3: Nanostorage). KW and EM acknowledge financial support from the National Science Center Poland (UMO-2016/23/G/ST5/04200).

Acknowledgments: The authors thank G. Kreutzer for technical support.

Conflicts of Interest: The authors declare no conflict of interest.

References

1. Palacín, M.R. Recent advances in rechargeable battery materials: A chemist's perspective. *Chem. Soc. Rev.* **2009**, *38*, 2565–2575. [CrossRef] [PubMed]
2. Aravindan, V.; Lee, Y.-S.; Madhavi, S. Research Progress on Negative Electrodes for Practical Li-Ion Batteries: Beyond Carbonaceous Anodes. *Adv. Energy Mater.* **2015**, *5*, 1402225. [CrossRef]
3. Mahmood, N.; Tang, T.; Hou, Y. Nanostructured Anode Materials for Lithium Ion Batteries: Progress, Challenge and Perspective. *Adv. Energy Mater.* **2016**, *6*, 1600374. [CrossRef]
4. Deng, Y.; Wan, L.; Xie, Y.; Qin, X.; Chen, G. Recent advances in Mn-based oxides as anode materials for lithium ion batteries. *RSC Adv.* **2014**, *4*, 23914–23935. [CrossRef]
5. Arico, A.S.; Bruce, P.; Scrosati, B.; Tarascon, J.-M.; van Schalkwijk, W. Nanostructured materials for advanced energy conversion and storage devices. *Nat. Mater.* **2005**, *4*, 366–377. [CrossRef]
6. Nazar, L.F.; Goward, G.; Leroux, F.; Duncan, M.; Huang, H.; Kerr, T.; Gaubicher, J. Nanostructured materials for energy storage. *Int. J. Inorg. Mater.* **2001**, *3*, 191–200. [CrossRef]
7. Armand, M.; Tarascon, J.-M. Building better batteries. *Nature* **2008**, *451*, 652–657. [CrossRef]
8. Goriparti, S.; Miele, E.; de Angelis, F.; Di Fabrizio, E.; Proietti Zaccaria, R.; Capiglia, C. Review on recent progress of nanostructured anode materials for Li-ion batteries. *J. Power Sources* **2014**, *257*, 421–443. [CrossRef]
9. Ji, L.; Lin, Z.; Alcoutlabi, M.; Zhang, X. Recent developments in nanostructured anode materials for rechargeable lithium-ion batteries. *Energy Environ. Sci.* **2011**, *4*, 2682–2699. [CrossRef]
10. Huang, X.; Cui, S.; Chang, J.; Hallac, P.B.; Fell, C.R.; Luo, Y.; Metz, B.; Jiang, J.; Hurley, P.T.; Chen, J. A hierarchical tin/carbon composite as an anode for lithium-ion batteries with a long cycle life. *Angew. Chem. Int. Ed. Engl.* **2015**, *54*, 1490–1493. [CrossRef]
11. Gao, G.; Le, Y.; Wu, H.B.; Lou, X.W.D. Hierarchical tubular structures constructed by carbon-coated α-Fe_2O_3 nanorods for highly reversible lithium storage. *Small* **2014**, *10*, 1741–1745. [CrossRef] [PubMed]
12. Wang, P.-P.; Sun, H.; Ji, Y.; Li, W.; Wang, X. Three-dimensional assembly of single-layered MoS(2). *Adv. Mater.* **2014**, *26*, 964–969. [CrossRef] [PubMed]
13. Li, Z.; Ottmann, A.; Zhang, T.; Sun, Q.; Meyer, H.-P.; Vaynzof, Y.; Xiang, J.; Klingeler, R. Preparation of hierarchical C@MoS_2@C sandwiched hollow spheres for lithium ion batteries. *J. Mater. Chem. A* **2017**, *5*, 3987–3994. [CrossRef]
14. Li, Z.; Ottmann, A.; Sun, Q.; Kast, A.K.; Wang, K.; Zhang, T.; Meyer, H.-P.; Backes, C.; Kübel, C.; Schröder, R.R.; et al. Hierarchical MoS 2 –carbon porous nanorods towards atomic interfacial engineering for high-performance lithium storage. *J. Mater. Chem. A* **2019**, *7*, 7553–7564. [CrossRef]
15. Lou, X.W.; Chen, J.S.; Chen, P.; Archer, L.A. One-Pot Synthesis of Carbon-Coated SnO_2 Nanocolloids with Improved Reversible Lithium Storage Properties. *Chem. Mater.* **2009**, *21*, 2868–2874. [CrossRef]
16. Kang, T.-W.; Lim, H.-S.; Park, S.-J.; Sun, Y.-K.; Suh, K.-D. Fabrication of flower-like tin/carbon composite microspheres as long-lasting anode materials for lithium ion batteries. *Mater. Chem. Phys.* **2017**, *185*, 6–13. [CrossRef]
17. Chen, X.; Xiao, T.; Wang, S.; Li, J.; Xiang, P.; Jiang, L.; Tan, X. Superior Li-ion storage performance of graphene decorated NiO nanowalls on Ni as anode for lithium ion batteries. *Mater. Chem. Phys.* **2019**, *222*, 31–36. [CrossRef]
18. Sun, H.; Sun, X.; Hu, T.; Yu, M.; Lu, F.; Lian, J. Graphene-Wrapped Mesoporous Cobalt Oxide Hollow Spheres Anode for High-Rate and Long-Life Lithium Ion Batteries. *J. Phys. Chem. C* **2014**, *118*, 2263–2272. [CrossRef]
19. Choi, S.H.; Ko, Y.N.; Jung, K.Y.; Kang, Y.C. Macroporous Fe_3O_4/carbon composite microspheres with a short Li+ diffusion pathway for the fast charge/discharge of lithium ion batteries. *Chemistry* **2014**, *20*, 11078–11083. [CrossRef]
20. Lei, C.; Han, F.; Sun, Q.; Li, W.-C.; Lu, A.-H. Confined nanospace pyrolysis for the fabrication of coaxial Fe_3O_4@C hollow particles with a penetrated mesochannel as a superior anode for Li-ion batteries. *Chemistry* **2014**, *20*, 139–145. [CrossRef]

21. Wu, C.; Li, X.; Li, W.; Li, B.; Wang, Y.; Wang, Y.; Xu, M.; Xing, L. Fe_2O_3 nanorods/carbon nanofibers composite: Preparation and performance as anode of high rate lithium ion battery. *J. Power Sources* **2014**, *251*, 85–91. [CrossRef]
22. Zhang, S.; Zhu, L.; Song, H.; Chen, X.; Zhou, J. Enhanced electrochemical performance of MnO nanowire/ graphene composite during cycling as the anode material for lithium-ion batteries. *Nano Energy* **2014**, *10*, 172–180. [CrossRef]
23. Shi, X.; Zhang, S.; Chen, X.; Tang, T.; Klingeler, R.; Mijowska, E. Ultrathin NiO confined within hollow carbon sphere for efficient electrochemical energy storage. *J. Alloys Compd.* **2019**, *797*, 702–709. [CrossRef]
24. Kan, J.; Wang, Y. Large and fast reversible Li-ion storages in Fe_2O_3-graphene sheet-on-sheet sandwich-like nanocomposites. *Sci. Rep.* **2013**, *3*, 3502. [CrossRef]
25. Park, S.-K.; Jin, A.; Yu, S.-H.; Ha, J.; Jang, B.; Bong, S.; Woo, S.; Sung, Y.-E.; Piao, Y. In Situ Hydrothermal Synthesis of Mn_3O_4 Nanoparticles on Nitrogen-doped Graphene as High-Performance Anode materials for Lithium Ion Batteries. *Electrochim. Acta* **2014**, *120*, 452–459. [CrossRef]
26. Zakharova, G.S.; Ottmann, A.; Möller, L.; Andreikov, E.I.; Fattakhova, Z.A.; Puzyrev, I.S.; Zhu, Q.; Thauer, E.; Klingeler, R. TiO_2/C nanocomposites prepared by thermal annealing of titanium glycerolate as anode materials for lithium-ion batteries. *J. Mater. Sci.* **2018**, *53*. [CrossRef]
27. Wenelska, K.; Ottmann, A.; Schneider, P.; Thauer, E.; Klingeler, R.; Mijowska, E. Hollow carbon sphere/metal oxide nanocomposites anodes for lithium-ion batteries. *Energy* **2016**, *103*, 100–106. [CrossRef]
28. Guan, X.; Nai, J.; Zhang, Y.; Wang, P.; Yang, J.; Zheng, L.; Zhang, J.; Guo, L. CoO Hollow Cube/Reduced Graphene Oxide Composites with Enhanced Lithium Storage Capability. *Chem. Mater.* **2014**, *26*, 5958–5964. [CrossRef]
29. Avogadro: An Open-Source Molecular Builder and Visualization Tool, 2.0. Available online: http://avogadro.cc/ (accessed on 18 January 2019).
30. Ottmann, A. Nanostrukturierte Kohlenstoff-Komposite und Ammoniumvanadate als Elektrodenmaterialien für Lithium-Ionen-Batterien. Ph.D. Thesis, Ruprecht-Karls-Universität, Heidelberg, Germany, 2018.
31. Dai, H. Carbon nanotubes: Opportunities and challenges. *Surf. Sci.* **2002**, *500*, 218–241. [CrossRef]
32. Wenelska, K.; Neef, C.; Schlestein, L.; Klingeler, R.; Kalenczuk, R.J.; Mijowska, E. Carbon nanotubes decorated by mesoporous cobalt oxide as electrode material for lithium-ion batteries. *Chem. Phys. Lett.* **2015**, *635*, 185–189. [CrossRef]
33. Zhang, X.; Zhou, Y.; Mao, Y.; Wei, M.; Chu, W.; Huang, K. Rapid synthesis of ultrafine $NiCo_2O_4$ nanoparticles loaded carbon nanotubes for lithium ion battery anode materials. *Chem. Phys. Lett.* **2019**, *715*, 278–283. [CrossRef]
34. Wen, Z.; Ci, S.; Mao, S.; Cui, S.; He, Z.; Chen, J. CNT@TiO_2 nanohybrids for high-performance anode of lithium-ion batteries. *Nanoscale Res. Lett.* **2013**, *8*, 499. [CrossRef] [PubMed]
35. Bhaskar, A.; Deepa, M.; Narasinga Rao, T. MoO_2/multiwalled carbon nanotubes (MWCNT) hybrid for use as a Li-ion battery anode. *ACS Appl. Mater. Interfaces* **2013**, *5*, 2555–2566. [CrossRef] [PubMed]
36. Poizot, P.; Laruelle, S.; Grugeon, S.; Dupont, L.; Tarascon, J.M. Nano-sized transition-metal oxides as negative-electrode materials for lithium-ion batteries. *Nature* **2000**, *407*, 496–499. [CrossRef] [PubMed]
37. Winter, M.; Besenhard, J.O.; Spahr, M.E.; Novák, P. Insertion Electrode Materials for Rechargeable Lithium Batteries. *Adv. Mater.* **1998**, *10*, 725–763. [CrossRef]
38. Tsang, S.C.; Chen, Y.K.; Harris, P.J.F.; Green, M.L.H. A simple chemical method of opening and filling carbon nanotubes. *Nature* **1994**, 159–162. [CrossRef]
39. Gellesch, M.; Dimitrakopoulou, M.; Scholz, M.; Blum, C.G.F.; Schulze, M.; van den Brink, J.; Hampel, S.; Wurmehl, S.; Büchner, B. Facile Nanotube-Assisted Synthesis of Ternary Intermetallic Nanocrystals of the Ferromagnetic Heusler Phase Co_2FeGa: Supporting Information. *Cryst. Growth Des.* **2013**, *13*, 2707–2710. [CrossRef]
40. Al Khabouri, S.; Al Harthi, S.; Maekawa, T.; Nagaoka, Y.; Elzain, M.E.; Al Hinai, A.; Al-Rawas, A.D.; Gismelseed, A.M.; Yousif, A.A. Composition, Electronic and Magnetic Investigation of the Encapsulated $ZnFe_2O_4$ Nanoparticles in Multiwall Carbon Nanotubes Containing Ni Residuals. *Nanoscale Res. Lett.* **2015**, *10*, 262. [CrossRef]
41. Ghunaim, R.; Damm, C.; Wolf, D.; Lubk, A.; Büchner, B.; Mertig, M.; Hampel, S. $Fe_{1-x}Ni_x$ Alloy Nanoparticles Encapsulated Inside Carbon Nanotubes: Controlled Synthesis, Structure and Magnetic Properties. *Nanomaterials* **2018**, *8*, 576. [CrossRef]

42. Ghunaim, R.; Eckert, V.; Scholz, M.; Gellesch, M.; Wurmehl, S.; Damm, C.; Büchner, B.; Mertig, M.; Hampel, S. Carbon nanotube-assisted synthesis of ferromagnetic Heusler nanoparticles of Fe 3 Ga (Nano-Galfenol). *J. Mater. Chem. C* **2018**, *6*, 1255–1263. [CrossRef]
43. Ghunaim, R.; Scholz, M.; Damm, C.; Rellinghaus, B.; Klingeler, R.; Büchner, B.; Mertig, M.; Hampel, S. Single-crystalline FeCo nanoparticle-filled carbon nanotubes: Synthesis, structural characterization and magnetic properties. *Beilstein J. Nanotechnol.* **2018**, *9*, 1024–1034. [CrossRef]
44. Haft, M.; Grönke, M.; Gellesch, M.; Wurmehl, S.; Büchner, B.; Mertig, M.; Hampel, S. Tailored nanoparticles and wires of Sn, Ge and Pb inside carbon nanotubes. *Carbon* **2016**, *101*, 352–360. [CrossRef]
45. Ottmann, A.; Scholz, M.; Haft, M.; Thauer, E.; Schneider, P.; Gellesch, M.; Nowka, C.; Wurmehl, S.; Hampel, S.; Klingeler, R. Electrochemical Magnetization Switching and Energy Storage in Manganese Oxide filled Carbon Nanotubes. *Sci. Rep.* **2017**, *7*, 13625. [CrossRef] [PubMed]
46. Jarosch, D. Crystal structure refinement and reflectance measurements of hausmannite, Mn_3O_4. *Mineral. Petrol.* **1987**, *37*, 15–23. [CrossRef]
47. Ferreira, T.A.S.; Waerenborgh, J.C.; Mendonça, M.H.R.M.; Nunes, M.R.; Costa, F.M. Structural and morphological characterization of $FeCo_2O_4$ and $CoFe_2O_4$ spinels prepared by a coprecipitation method. *Solid State Sci.* **2003**, *5*, 383–392. [CrossRef]
48. Kapoor, A.; Singh, N.; Dey, A.B.; Nigam, A.K.; Bajpai, A. 3d transition metals and oxides within carbon nanotubes by co-pyrolysis of metallocene & camphor: High filling efficiency and self-organized structures. *Carbon* **2018**, *132*, 733–745. [CrossRef]
49. Heider, R. Nanoskalige Sn-Co-Verbindungen Durch Füllen von CNT. Master Thesis, BTU Cottbus-Senftenberg, Cottbus, Germany, 2015.
50. Frackowiak, E.; Béguin, F. Electrochemical storage of energy in carbon nanotubes and nanostructured carbons. *Carbon* **2002**, *40*, 1775–1787. [CrossRef]
51. Chew, S.Y.; Ng, S.H.; Wang, J.; Novák, P.; Krumeich, F.; Chou, S.L.; Chen, J.; Liu, H.K. Flexible free-standing carbon nanotube films for model lithium-ion batteries. *Carbon* **2009**, *47*, 2976–2983. [CrossRef]
52. Xiong, Z.; Yun, Y.; Jin, H.-J. Applications of Carbon Nanotubes for Lithium Ion Battery Anodes. *Materials* **2013**, *6*, 1138–1158. [CrossRef]
53. Fang, X.; Lu, X.; Guo, X.; Mao, Y.; Hu, Y.-S.; Wang, J.; Wang, Z.; Wu, F.; Liu, H.; Chen, L. Electrode reactions of manganese oxides for secondary lithium batteries. *Electrochem. Commun.* **2010**, *12*, 1520–1523. [CrossRef]
54. Zhong, K.; Xia, X.; Zhang, B.; Li, H.; Wang, Z.; Chen, L. MnO powder as anode active materials for lithium ion batteries. *J. Power Sources* **2010**, *195*, 3300–3308. [CrossRef]
55. Tyler, R.W. The Magnetic Susceptibility of MnO as a Function of the Temperature. *Phys. Rev.* **1933**, *44*, 776–777. [CrossRef]
56. Bai, Z.; Zhang, X.; Zhang, Y.; Guo, C.; Tang, B. Facile synthesis of mesoporous Mn_3O_4 nanorods as a promising anode material for high performance lithium-ion batteries. *J. Mater. Chem. A* **2014**, *2*, 16755–16760. [CrossRef]
57. Li, L.; Guo, Z.; Du, A.; Liu, H. Rapid microwave-assisted synthesis of Mn_3O_4–graphene nanocomposite and its lithium storage properties. *J. Mater. Chem.* **2012**, *22*, 3600–3605. [CrossRef]
58. Wang, Z.-H.; Yuan, L.-X.; Shao, Q.-G.; Huang, F.; Huang, Y.-H. Mn_3O_4 nanocrystals anchored on multi-walled carbon nanotubes as high-performance anode materials for lithium-ion batteries. *Mater. Lett.* **2012**, *80*, 110–113. [CrossRef]
59. Luo, S.; Wu, H.; Wu, Y.; Jiang, K.; Wang, J.; Fan, S. Mn_3O_4 nanoparticles anchored on continuous carbon nanotube network as superior anodes for lithium ion batteries. *J. Power Sources* **2014**, *249*, 463–469. [CrossRef]
60. Lavela, P.; Ortiz, G.F.; Tirado, J.L.; Zhecheva, E.; Stoyanova, R.; Ivanova, S. High-Performance Transition Metal Mixed Oxides in Conversion Electrodes: A Combined Spectroscopic and Electrochemical Study. *J. Phys. Chem. C* **2007**, *111*, 14238–14246. [CrossRef]
61. De las Casas, C.; Li, W. A review of application of carbon nanotubes for lithium ion battery anode material. *J. Power Sources* **2012**, *208*, 74–85. [CrossRef]
62. Varzi, A.; Täubert, C.; Wohlfahrt-Mehrens, M.; Kreis, M.; Schütz, W. Study of multi-walled carbon nanotubes for lithium-ion battery electrodes. *J. Power Sources* **2011**, *196*, 3303–3309. [CrossRef]
63. Chu, Y.-Q.; Fu, Z.-W.; Qin, Q.-Z. Cobalt ferrite thin films as anode material for lithium ion batteries. *Electrochim. Acta* **2004**, *49*, 4915–4921. [CrossRef]

64. Verma, P.; Maire, P.; Novák, P. A review of the features and analyses of the solid electrolyte interphase in Li-ion batteries. *Electrochim. Acta* **2010**, *55*, 6332–6341. [CrossRef]
65. Gnanamuthu, R.; Lee, C.W. Electrochemical properties of Super P carbon black as an anode active material for lithium-ion batteries. *Mater. Chem. Phys.* **2011**, *130*, 831–834. [CrossRef]
66. Cabana, J.; Monconduit, L.; Larcher, D.; Palacín, M.R. Beyond intercalation-based Li-ion batteries: The state of the art and challenges of electrode materials reacting through conversion reactions. *Adv. Mater.* **2010**, *22*, E170–E192. [CrossRef] [PubMed]
67. Wang, Y.; Su, D.; Ung, A.; Ahn, J.H.; Wang, G. Hollow $CoFe_2O_4$ nanospheres as a high capacity anode material for lithium ion batteries. *Nanotechnology* **2012**, *23*, 55402. [CrossRef] [PubMed]
68. Wu, L.; Xiao, Q.; Li, Z.; Lei, G.; Zhang, P.; Wang, L. $CoFe_2O_4$/C composite fibers as anode materials for lithium-ion batteries with stable and high electrochemical performance. *Solid State Ion.* **2012**, *215*, 24–28. [CrossRef]
69. Wang, L.; Zhuo, L.; Cheng, H.; Zhang, C.; Zhao, F. Porous carbon nanotubes decorated with nanosized cobalt ferrite as anode materials for high-performance lithium-ion batteries. *J. Power Sources* **2015**, *283*, 289–299. [CrossRef]
70. Zhang, Z.; Wang, Y.; Zhang, M.; Tan, Q.; Lv, X.; Zhong, Z.; Su, F. Mesoporous $CoFe_2O_4$ nanospheres cross-linked by carbon nanotubes as high-performance anodes for lithium-ion batteries. *J. Mater. Chem. A* **2013**, *1*, 7444–7450. [CrossRef]
71. Ren, S.; Zhao, X.; Chen, R.; Fichtner, M. A facile synthesis of encapsulated $CoFe_2O_4$ into carbon nanofibres and its application as conversion anodes for lithium ion batteries. *J. Power Sources* **2014**, *260*, 205–210. [CrossRef]
72. Zhang, L.; Wei, T.; Jiang, Z.; Liu, C.; Jiang, H.; Chang, J.; Sheng, L.; Zhou, Q.; Yuan, L.; Fan, Z. Electrostatic interaction in electrospun nanofibers: Double-layer carbon protection of $CoFe_2O_4$ nanosheets enabling ultralong-life and ultrahigh-rate lithium ion storage. *Nano Energy* **2018**, *48*, 238–247. [CrossRef]
73. Laruelle, S.; Grugeon, S.; Poizot, P.; Dollé, M.; Dupont, L.; Tarascon, J.-M. On the Origin of the Extra Electrochemical Capacity Displayed by MO/Li Cells at Low Potential. *J. Electrochem. Soc.* **2002**, *149*, A627–A634. [CrossRef]
74. Jamnik, J.; Maier, J. Nanocrystallinity effects in lithium battery materials. *Phys. Chem. Chem. Phys.* **2003**, *5*, 5215. [CrossRef]
75. Larcher, D.; Bonnin, D.; Cortes, R.; Rivals, I.; Personnaz, L.; Tarascon, J.-M. Combined XRD, EXAFS, and Mössbauer Studies of the Reduction by Lithium of α-Fe_2O_3 with Various Particle Sizes. *J. Electrochem. Soc.* **2003**, *150*, A1643–A1650. [CrossRef]
76. Larcher, D.; Masquelier, C.; Bonnin, D.; Chabre, Y.; Masson, V.; Leriche, J.-B.; Tarascon, J.-M. Effect of Particle Size on Lithium Intercalation into α-Fe_2O_3. *J. Electrochem. Soc.* **2003**, *150*, A133–A139. [CrossRef]
77. Morales, J.; Sánchez, L.; Martín, F.; Berry, F.; Ren, X. Synthesis and Characterization of Nanometric Iron and Iron-Titanium Oxides by Mechanical Milling. *J. Electrochem. Soc.* **2005**, *152*, A1748–A1754. [CrossRef]
78. Cherian, C.T.; Sundaramurthy, J.; Kalaivani, M.; Ragupathy, P.; Kumar, P.S.; Thavasi, V.; Reddy, M.V.; Sow, C.H.; Mhaisalkar, S.G.; Ramakrishna, S.; et al. Electrospun α-Fe_2O_3 nanorods as a stable, high capacity anode material for Li-ion batteries. *J. Mater. Chem.* **2012**, *22*, 12198–12204. [CrossRef]
79. Hariharan, S.; Saravanan, K.; Balaya, P. Lithium Storage Using Conversion Reaction in Maghemite and Hematite. *Electrochem. Solid State Lett.* **2010**, *13*, A132–A134. [CrossRef]
80. Thackeray, M.M. Spinel Electrodes for Lithium Batteries. *J. Am. Ceram. Soc.* **1999**, *82*, 3347–3354. [CrossRef]
81. Thackeray, M.M.; David, W.I.F.; Goodenough, J.B. Structural characterization of the lithiated iron oxides $Li_xFe_3O_4$ and $Li_xFe_2O_3$ (0<x<2). *Mater. Res. Bull.* **1982**, *17*, 785–793. [CrossRef]
82. Xu, J.-S.; Zhu, Y.-J. Monodisperse Fe_3O_4 and γ-Fe_2O_3 magnetic mesoporous microspheres as anode materials for lithium-ion batteries. *ACS Appl. Mater. Interfaces* **2012**, *4*, 4752–4757. [CrossRef]
83. Yuan, S.; Zhou, Z.; Li, G. Structural evolution from mesoporous α-Fe_2O_3 to Fe_3O_4@C and γ-Fe_2O_3 nanospheres and their lithium storage performances. *CrystEngComm* **2011**, *13*, 4709–4713. [CrossRef]
84. Yan, N.; Zhou, X.; Li, Y.; Wang, F.; Zhong, H.; Wang, H.; Chen, Q. Fe_2O_3 Nanoparticles Wrapped in Multi-walled Carbon Nanotubes With Enhanced Lithium Storage Capability. *Sci. Rep.* **2013**, *3*, 3392. [CrossRef] [PubMed]
85. Zhuo, L.; Wu, Y.; Ming, J.; Wang, L.; Yu, Y.; Zhang, X.; Zhao, F. Facile synthesis of a Co_3O_4–carbon nanotube composite and its superior performance as an anode material for Li-ion batteries. *J. Mater. Chem. A* **2013**, *1*, 1141–1147. [CrossRef]

86. Xu, M.; Wang, F.; Zhang, Y.; Yang, S.; Zhao, M.; Song, X. Co$_3$O$_4$-carbon nanotube heterostructures with bead-on-string architecture for enhanced lithium storage performance. *Nanoscale* **2013**, *5*, 8067–8072. [CrossRef] [PubMed]
87. Obrovac, M.N.; Chevrier, V.L. Alloy negative electrodes for Li-ion batteries. *Chem. Rev.* **2014**, *114*, 11444–11502. [CrossRef]
88. Wu, S.; Han, C.; Iocozzia, J.; Lu, M.; Ge, R.; Xu, R.; Lin, Z. Germanium-Based Nanomaterials for Rechargeable Batteries. *Angew. Chem. Int. Ed.* **2016**, *55*, 7898–7922. [CrossRef] [PubMed]
89. Srajer, G.; Lewis, L.H.; Bader, S.D.; Epstein, A.J.; Fadley, C.S.; Fullerton, E.E.; Hoffmann, A.; Kortright, J.B.; Krishnan, K.M.; Majetich, S.A.; et al. Advances in nanomagnetism via X-ray techniques. *J. Magn. Magn. Mater.* **2006**, *307*, 1–31. [CrossRef]
90. Goward, G.R.; Taylor, N.J.; Souza, D.C.S.; Nazar, L.F. The true crystal structure of Li$_{17}$M$_4$ (M=Ge, Sn, Pb)—revised from Li$_{22}$M$_5$. *J. Alloys Compd.* **2001**, *329*, 82–91. [CrossRef]
91. Yin, F.; Su, X.; Li, Z.; Wang, J. Thermodynamic assessment of the Li–Sn (Lithium–Tin) system. *J. Alloys Compd.* **2005**, *393*, 105–108. [CrossRef]
92. Todd, A.D.W.; Mar, R.E.; Dahn, J.R. Combinatorial Study of Tin-Transition Metal Alloys as Negative Electrodes for Lithium-Ion Batteries. *J. Electrochem. Soc.* **2006**, *153*, A1998–A2005. [CrossRef]
93. Zhang, J.-j.; Xia, Y.-y. Co-Sn Alloys as Negative Electrode Materials for Rechargeable Lithium Batteries. *J. Electrochem. Soc.* **2006**, *153*, A1466–A1471. [CrossRef]
94. Courtney, I.A.; Dahn, J.R. Electrochemical and In Situ X-Ray Diffraction Studies of the Reaction of Lithium with Tin Oxide Composites. *J. Electrochem. Soc.* **1997**, *144*, 2045–2052. [CrossRef]
95. Ionica-Bousquet, C.M.; Lippens, P.E.; Aldon, L.; Olivier-Fourcade, J.; Jumas, J.C. In situ 119Sn Mössbauer Effect Study of Li–CoSn$_2$ Electrochemical System. *Chem. Mater.* **2006**, *18*, 6442–6447. [CrossRef]
96. Wang, Y.; Wu, M.; Jiao, Z.; Lee, J.Y. Sn@CNT and Sn@C@CNT nanostructures for superior reversible lithium ion storage. *Chem. Mater.* **2009**, *21*, 3210–3215. [CrossRef]
97. Zakharova, G.S.; Thauer, E.; Wegener, S.A.; Nölke, J.-H.; Zhu, Q.; Klingeler, R. Hydrothermal microwave-assisted synthesis of Li$_3$VO$_4$ as an anode for lithium-ion battery. *J. Solid State Electrochem.* **2019**, *23*, 2205–2212. [CrossRef]

© 2020 by the authors. Licensee MDPI, Basel, Switzerland. This article is an open access article distributed under the terms and conditions of the Creative Commons Attribution (CC BY) license (http://creativecommons.org/licenses/by/4.0/).

Review

Functionalized Carbon Nanostructures Versus Drug Resistance: Promising Scenarios in Cancer Treatment

Manuela Curcio [1], Annafranca Farfalla [1], Federica Saletta [2,3,4], Emanuele Valli [2,3], Elvira Pantuso [1], Fiore Pasquale Nicoletta [1], Francesca Iemma [1], Orazio Vittorio [2,3,4,*,†] and Giuseppe Cirillo [1,*,†]

1. Department of Pharmacy, Health and Nutritional Sciences, University of Calabria, 87036 Rende (CS), Italy; manuela.curcio@unical.it (M.C.); annafranca.farfalla@gmail.com (A.F.); elvirapnt.ep@gmail.com (E.P.); fiore.nicoletta@unical.it (F.P.N.); francesca.iemma@unical.it (F.I.)
2. Lowy Cancer Research Centre, Children's Cancer Institute, UNSW Sydney, NSW 2031, Australia; FSaletta@ccia.org.au (F.S.); EValli@ccia.org.au (E.V.)
3. School of Women's and Children's Health, Faculty of Medicine, UNSW Sydney, NSW 2052, Australia
4. ARC Centre of Excellence for Convergent BioNano Science and Technology, Australian Centre for NanoMedicine, UNSW Sydney, NSW 2052, Australia
* Correspondence: OVittorio@ccia.unsw.edu.au (O.V.); giuseppe.cirillo@unical.it (G.C.); Tel.: +61-2938-51557 (O.V.); +39-0984-493011 (G.C.)
† These authors contributed equally to this work.

Academic Editor: Filippo Parisi
Received: 26 March 2020; Accepted: 28 April 2020; Published: 30 April 2020

Abstract: Carbon nanostructures (CN) are emerging valuable materials for the assembly of highly engineered multifunctional nanovehicles for cancer therapy, in particular for counteracting the insurgence of multi-drug resistance (MDR). In this regard, carbon nanotubes (CNT), graphene oxide (GO), and fullerenes (F) have been proposed as promising materials due to their superior physical, chemical, and biological features. The possibility to easily modify their surface, conferring tailored properties, allows different CN derivatives to be synthesized. Although many studies have explored this topic, a comprehensive review evaluating the beneficial use of functionalized CNT vs G or F is still missing. Within this paper, the most relevant examples of CN-based nanosystems proposed for MDR reversal are reviewed, taking into consideration the functionalization routes, as well as the biological mechanisms involved and the possible toxicity concerns. The main aim is to understand which functional CN represents the most promising strategy to be further investigated for overcoming MDR in cancer.

Keywords: carbon nanostructures; carbon nanohybrids; cancer therapy; multi-drug resistance

1. Introduction

The USA National Cancer Institute defines cancer as several diseases characterized by an uncontrollable proliferation of abnormal cells invading surrounding tissues [1], with over 10 million new cases diagnosed each year with a survival rate of around 40% [2]. The high incidence of unfavorable prognoses is related to a multitude of factors, including late stage diagnosis, cancer cell plasticity, lack of therapeutic approaches for eradicating disseminated cancer cells, and development of multi-drug resistance (MDR) [3]. On the basis of the underlying developing mechanism, MDR can be classified as intrinsic and extrinsic [4], if depends on acquired genetic alterations in tumor cells [5] or to prolonged exposure to chemotherapy [6], respectively.

The main extrinsic MDR mechanisms involve the reduction of either intracellular drug concentration or activity, and the alteration of cellular apoptotic pathways [7]. More in details,

MDR is characterized by the variation of either cell membranes (protein and lipid composition) or cytoplasm (intracellular endocytic vesicles) and nuclei (genetic machinery) features [8].

Overcoming MDR is one of the big challenges to ensure the success of chemotherapy, with nanotechnology offering powerful tools for addressing this issue. Different nanomaterials, including metal-based (e.g., iron, silver, and gold nanoparticles), carbon-based (e.g., carbon nanotubes, graphene), and polymer-based (e.g., polymer therapeutics and polymer nanoparticles) materials, have been proposed for various aspects of cancer therapy. Such materials, often combined in composite systems, have gained intense research interest due to their ability to enhance the therapeutic effectiveness and reduce the systemic side effects of conventional cytotoxic drugs [9] (Figure 1).

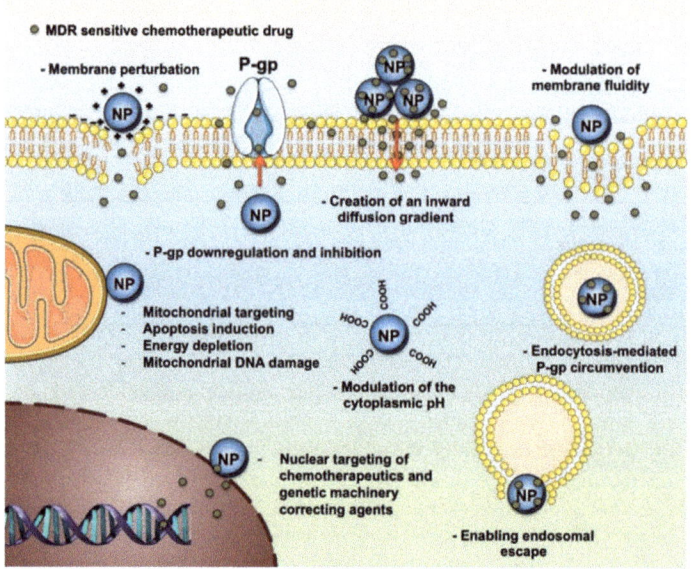

Figure 1. General representation of the main MDR reversal mechanisms by nanoparticle systems (NP). Reproduced with permission from [4]. Copyright Elsevier (2017)

Carbon nanostructures (CN) consist of sp^2 carbon atoms with different spatial arrangements, mainly consisting in fullerenes (F, 0D) [10], carbon nanotubes (CNT, 1D) [11], graphene (G, 2D) [12], and graphite/diamond (3D) [13].

G, a flat honeycomb lattice composed of a single layer of hexagonal carbon atoms held together by a backbone of overlapping sp^2 hybrid bonds, can be assumed as the basic building block for other CN [14].

F are irregular stacked graphene sheets arranged in hollow spherical or ellipsoid structures, where carbon atoms form hexagonal or pentagonal rings. Various forms of fullerenes have been found, including C_{60}, C_{70}, C_{76}, C_{80}, C_{84}, with sizes ranging from 30 up to 3000 carbon atoms [15]. The most stable fullerene is C_{60}, which is composed of 12 pentagonal and 20 hexagonal carbon atom rings.

CNTs can be imaginatively produced by rolling up one single-walled (SWCNT) or multi-walled CNTs (MWCNT) layer of graphene sheet to form cylindrical tubes with a pore diameter <100 nm and a length on the micron scale [16], being closed at the ends with fullerenes halfspheres [17].

Other peculiar classes of CN include quasi spherical graphene structures (graphene quantum dots—GQD) [18], elongated strips of graphene (carbon nanoribbons) [19], rolled graphene sheets with a closed horn-shaped tip (carbon nanohorns) [20], conical cap curved by pentagonal carbon rings (carbon nanocones) [21].

Here, we are giving an overview of the most relevant vehicles based on the CN mainly proposed for the triggered delivery of chemotherapeutics to resistant cancer cells, taking into consideration the biological performances as well as the technological features of the delivery systems. We aim to show how this recent field of research, has contributed to improve the knowledge for developing effective cancer treatments, and what are the potential directions where researchers can focus their future studies to accelerate the translation of their discoveries to clinical trials. By highlighting the strength and the weakness of each CN and derivatization strategy, we aim to show the scientific evidences supporting the hypothesis that the use of G instead of CNT derivatives can address the toxicity concerns and open new perspectives for improving drug efficiency in fighting cancer.

2. Carbon Nanostructures and Cancer: Toxicity Concerns and Needs for Tailored Functionalization

CN have aroused great interest among the scientific community for a plethora of applications [22], in physics, chemistry, material science, engineering, electronics, biology, and medicine [23,24] by virtue of their peculiar properties, such as high stability, electrical conductivity, versatile derivatization routes, NIR absorption, and ability to easily penetrate cell membranes. In the biomedical field, CN (CNT, G and C_{60} in particular) represent valuable tools for bio-sensing and bio-imaging [25], as well as for the design of tailored drug carrier systems [26,27]. Their high affinity for organic molecules (independent of molecular weight) confers high drug loading ability, allowing drug stabilization, and improved pharmacokinetic profiles, resulting in enhanced cell uptake [28]. Moreover, when combined with proper materials, CN can modulate the release of a therapeutic agent upon either the application of external stimuli such as magnetic and electric [29,30] fields, or the variation of environmental parameters such pH [31] and temperature [32]. Thus, CN are widely exploited as platforms for delivering bioactive molecules and genes with improved efficiency.

On the other hands, a preliminary functionalization of CN is required, since pristine materials suffer from severe toxic effects due to the lack of solubility in physiological environment. The toxicity of CN mainly arises from the interference with the cell membrane integrity and function, the damage of DNA and RNA, as well as the induction of oxidative stress, inflammatory response, apoptosis, autophagy, and necrosis [33]. In detail, the different CN show specific toxicity profiles, which are relative to their peculiar morphological and shape features, which are affecting cell response to drugs and thus the potential undesired side-effects.

According to the World Health Organisation (WHO), the CNT toxicity concerns come from their fiber-like structure and similarity to asbestos fibers, with carcinogenic effects mainly related to their size [33]. Research data have demonstrated that intra-abdominal injection of long MWCNT determines chronic inflammation of the abdominal wall with the formation of mesothelioma [34]. On the contrary, no inflammation was detected in the case of short MWCNT due to the complete phagocytosis by macrophages, although activated phagocytes can result in the generation of Reactive Oxygen Species (ROS) and thus in late DNA damage [35]. Minimal genotoxicity was recorded upon exposure to low F doses as a consequence of the photosensitizing effect induced by ROS generation, while inflammation was detected at high dosages due to nitric oxide synthase-dependent induction of cyclooxygenase-2 [33]. The high rigidity of pristine G may induce incomplete phagocytosis and thus inflammation and ROS generation, although it is not totally clarified if ROS originate from the its surface or is formed by cellular reactions involving mitochondria and leukocyte [36].

Different approaches have been proposed to improve the water affinity and thus the biocompatibility of CN via the modification of their surface properties [37,38]. The functionalization routes can be divided in two main categories: (i) the covalent attachment of chemical functionalities with change (from sp^2 to sp^3) in the hybridization of carbon atoms in the site of reaction [39], and (ii) the noncovalent adsorption/wrapping (π-π stacking) of tailored functional molecules via hydrophobic, electrostatic and Van der Waals interactions, without any chemical changes in the electron patterns of the CN surface [40–42].

Different strategies for the covalent functionalization of CN, involving the typical reactivity of the sp^2 carbon atoms on their surface, are summarized in Table 1.

Each approach shows peculiar features in terms of chemical compatibility, reaction conditions, and derivatization degree [43,44], which can be finely tuned according to the specific application needs (Figure 2).

Table 1. Main CN covalent functionalization routes.

	Reaction			Ref	
N.	Type	Derivatizing Agents	CNT	G	C_{60}
1	Halogenation	F_2	[45]	[46]	[47]
2	Hydrogenation	H_2	[48]	[49]	[50]
3	Oxidation	a) HNO_3/H_2SO_4	[51]	[52]	[53]
		b) H_2O_2	[54]	[55]	[56]
		c) O_3	[57]	[58]	[59]
4	Nucleophilic Addition	Nu^-	[60]	[61]	[62]
5	Radical Coupling	a) $R\text{-}Ar\text{-}N_2^+$	[63]	[64]	[65]
		b) $R\text{-}Ar\text{-}NH_2$	[66]	[67]	[68]
6	Electrophilic Addition	RCOX	[69]	[70]	[71]
7	Cycloaddition	a) $R_2C:$	[72]	[73]	[74]
		b) $N_3\text{-}COOR$	[75]	[76]	[77]
		c) $R\text{-}NHCH_2COOH/(CH_2O)_n$	[78]	[79]	[80]
		d) $EtOOCCH_2COOEt$	[81]	[82]	[83]
		e) $R\text{-}C=N\text{-}NH\text{-}Ar$	[84]	[85]	[86]
		f) $\text{-}C=C(R)\text{-}C(R)=C\text{-}$	[87]	[88]	[89]

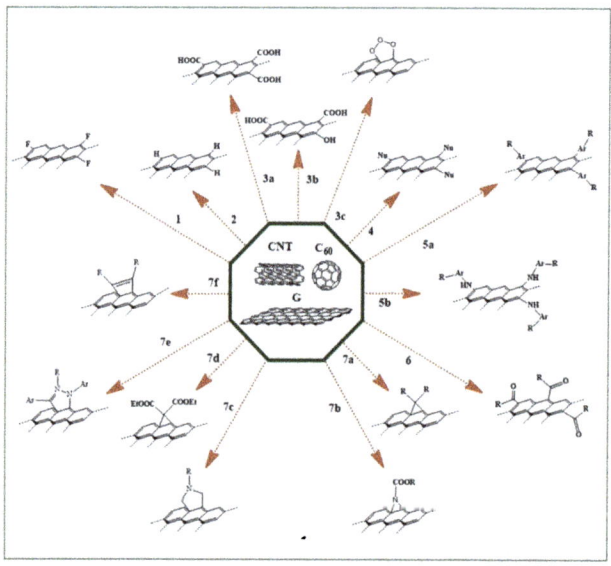

Figure 2. Schematic representation of the main carbon nanostructures covalent functionalization routes. CNT: carbon nanotubes; G: graphene; F: fullerenes. For each reaction, number codes and derivatizing agents are reported in Table 1.

In all cases, intermediate species with high chemical versatility are obtained, with further derivatization processes allowing the fabrication of functional materials for cancer therapy [90]. CNT-based nanosystems proposed for the effective treatment of drug-resistant cancer cells are mainly based on condensation reactions on oxidized CNT (oxCNT), prepared by chemical- [91], photo- [92] or gas- [93] based treatment. Depending on the reaction conditions and the exposure time, such methods are suitable for either the CNT purification from the residual impurities of the synthetic procedures or for the introduction of oxygen-rich functionalities (e.g., hydroxyl and carboxyl groups) on their surface [94] with an increase of water solubility up to 2 mg mL^{-1} [95]. It should be pointed out that, although CNT water dispersions form colloidal suspensions rather than true solutions in the molecular sense, the term "solution" is widely accepted when describing such dispersions [96]. The oxidation reaction starts from carbon bonds at the tips, which are characterized by a higher reactivity due to their larger curvature [97], although the hexagonal cylindrical tube walls are also involved with the formation of defects sites on CNT surface [16]. To avoid the formation of such defects, Prato and co-workers developed a [1,3] cycloaddition reaction of azomethine ylides generated in-situ by thermal condensation of aldehydes and α-amino acids [11]. Under mild reaction conditions, the azomethine ylides are coupled to π-bonds with the formation of pyrrolidine rings, allowing to finely modulate the chemical properties of the side chains on the final product [98]. The advantages of this approach are the simultaneous improvement of the CNT water affinity, the removal of the metal nanoparticles and amorphous carbon impurities, as well as the possibility to thermally remove the introduced organic groups, restoring the original CNT structure [99]. Another key CNT functionalization route is fluorination, defined as the breaking of the conjugated π layers on CNT surface with the formation of C-F bonds. Methods for fluorination mainly consist in the use of fluorine gas mixtures under proper temperature and positive pressure conditions [100] or plasma gas containing fluorine in vacuum [101]. The obtained materials possess a higher solubility in polar solvents (e.g., 1 mg mL^{-1} in alcohol) and are suitable for further derivatization processes by nucleophilic reactions [102].

Similarly to CNT, graphene oxide (GO) is the most used derivative of G. GO is obtained by exfoliation of graphite by treatment with strong oxidizing agents (e.g., Hummers' method) [103,104]. The exact structure of GO is difficult to determine, but evidences suggest that the sp^2-hybridized G lattice is interrupted by hydroxyl (−OH), epoxide (−O−), and carbonyl (−C=O) groups, while carboxyl (−COOH) groups are located at the edge [105]. The presence of aromatic network of sp^2-hybridized carbons and -COOH groups linked to sp^3-hybridized carbons is responsible for a good hydrophobic to hydrophilic balance of GO derivatives, allowing a high affinity for organic molecules, as well as a high water affinity and thus biocompatibility [106,107]. Furthermore, the oxygen-rich structures act as reactive sites for chemical functionalization with biocompatible and/or bioactive molecules [108].

rGO, a reduced form of GO with lower oxygen content [109], is another valuable G derivative with enhanced ability to load lipophilic species through π-π stacking [110,111]. The availability of different reducing agents and reaction conditions allows the extent of GO reduction and thus the drug-carrier interactions to be finely modulated [112]. Finally, graphene quantum dots (GQDs) are zero-dimensional G derivatives consisting in few layers of G sheets with size less than 20 nm [113]. They are attracting tremendous interest for the preparation of highly engineered carrier systems [114]. GDQ are obtained from large G sheets by chemical oxidation, thermal, ultrasound or oxygen plasma treatments [115].

3. Carbon Nanostructures Fighting Multi-Drug Resistance

The employment of nanocarriers, and of those based on CN in particular, have offered different solutions in cancer therapy [116], including the possibility to minimize, circumvent, or even reverse MDR [117] via two main effects:

a)   intrinsic MDR reversing properties [118];
b)   delivery of MDR reversing agents acting alone or in combination [119].

According to the biological pathways involved in the MDR reversal, the main mechanisms can be summarized as follows:

a) inhibition of drug efflux pumps;
b) increase of intracellular drug concentration and endosomal escape (enhanced uptake);
c) damage of cell membrane and/or intracellular organelles;
d) phototherapy.

In the next sections, we analyze the key examples of CN carrier systems proposed as MDR reversing methods, highlighting both the carrier features (e.g., used CN, preparation method, presence of anchored functional moieties, and/or targeting effect), and the MDR reversion route. Afterwards, the outcomes of the reviewed studies are summarized, and the efficiency of each system is compared respect to their own side effects.

3.1. MDR Reversal by Inhibition of Efflux Pumps

A key mechanism involved in either acquired or intrinsic MDR is the ability of integral membrane transporters to expel xenobiotic substances from living cells, including antibiotics and anti-cancer agents [120]. The P-glycoprotein (P-gp or MDR1), the multidrug resistance-associated protein 1 (MRP1), and the breast cancer resistance protein (BCRP) belong to the superfamily of ATP-binding cassette (ABC) transporters [121]. As the name suggests, these proteins transport a broad range of substrates across biological membranes against concentration gradients by using the energy of ATP hydrolysis [122]. P-gp is physiologically expressed in epithelial cells of the major excretory organs (e.g., liver, kidney, lung) and in the capillary endothelial cells at the blood–tissue interfaces, where plays a key role in preventing the entry and/or facilitating the elimination of drugs and toxins [123]. Moreover, since its overexpression in malignant cells' membrane is often associated with adverse prognosis, extensive efforts have been made for developing effective P-gp inhibitors able to antagonize its activity. Nanoparticle systems offer opportunities for P-gp inhibition by virtue of either their intrinsic inhibitory effect, or the possibility to co-deliver cytotoxic drugs and P-gp inhibitors [122]. In accordance with Wang et al.'s statement [124], since few experimental works have been designed to clearly determine the molecular mechanism at the basis of CN activity in MDR cancer cells, the molecular pathways involved in MDR reversal are an object of diffuse debate among the scientific community [125]. It can be hypothesized that CN enhance the drug uptake allowing an improved intracellular drug concentration, or that they directly inhibit P-gp activity. Experimental and theoretical methods demonstrated negligible C_{60} efflux by P-gp protein, which was confirmed with the absence of P-gp mediated efflux of a fluorescent substrate model [125]. Nevertheless, Shityakov and Föster [126] demonstrated the high affinity of C_{60} to P-gp, thus hypothesizing C_{60} ability to act as competitive substrate to be expelled from cells. In contrast, SWCNT, although possessing high affinity for the P-gp intracellular domains, cannot be effluxed because of unfavorable thermodynamics [126]. The CN employed for MDR reversal via P-gp modulation are summarized in Table 2.

Table 2. MDR reversal by CN via inhibition of efflux pumps.

Carrier			Delivery Properties			Cancer Model			Ref
CN	Derivatizing Agent	Bioactive Agent	DL	Responsivity	Tissue	In Vitro	In Vivo		
oxMWCNT	PEG-NH$_2$ Condensation	—	—	—	Cervix	HeLa	—	[127]
					Liver	HepG2		
						HepG2/R		
					Blood	K562		
						K562R		

Table 2. Cont.

Carrier		Delivery Properties			Cancer Model			Ref
CN	Derivatizing Agent	Bioactive Agent	DL	Responsivity	Tissue	In Vitro	In Vivo	
oxSWCNT	—	N-TAM-TEG *Condensation*		pH	Breast	MDA-MB-231/R	—	[128]
		Q π-π *Stacking*						
MWCNT	TCM *Coating*	—	—	—	Colon	Caco-2	—	[124]
GQD	—	DOX π-π *Stacking*	—	—	Breast	MCF-7	—	[129]
						MCF-7/ADR		
					Liver	SMMC-7721		
					Colon	Caco-2		
					Blood	HL-60		
SWCNT/ oxSWCNT/ MWCNT	—	—	—	—	Bone	MNNG/HOS	MNNG/ HOS	[130]

DL: drug loading % (*w/w*); DOX: doxorubicin; GQD: graphene quantum dots; MWCNT: multi-walled carbon nanotubes; oxMWCNT: oxidized MWCNT; PEG: polyethylene glycol; Q: quercetin; SWCNT: single-walled carbon nanotubes; oxSWCNT: oxidized SWCNT; N-TAM: N-desmethyltamoxifen; TCM: tissue culture medium; TEG: tetraethylene glycol.

Two key studies demonstrated the P-gp inhibitory activity of CNT loaded [127] or conjugated [128] with fluorescent probe and anticancer drug, respectively. In the first case, the authors showed that a higher efflux of the substrate was detected because of increased P-gp ATPase activity. On the other hand, the conjugation derivative was found to possess a P-gp inhibitory activity due to both CN and loaded P-gp inhibitor (Quercetin). These contradictory results clearly proved the need of further experiments to determine the mechanism of CN driven MDR reversal.

MDR reversal can also occur due to gene expression down-regulation. Wang et al. [124] showed that pristine MWCNT decreased the expression of proto-oncogene c-Myc, (involved in the regulation of ABC gene expression) without damaging cell membrane or inducing oxidative stress. Similarly, Luo et al. [129] reported the ability of GQDs to interact with P-gp C-rich regions (MRP1, and BCRP) resulting in a significant down-regulation of its expression and a significant reversal of Doxorubicin resistance in MCF-7/ADR breast cancer cells. Furthermore, Miao et al. [130] showed the ability of either pristine or oxidized single walled CNT (oxSWCNT) to inhibit cancer proliferation through reduction of tumor micro-vessel density and suppression of the TGFb1-signalling in osteosarcoma stem cells.

3.2. MDR Reversal by Enhanced Cellular Uptake

The high cell uptake capability of CN can be exploited in the attempt to revert MDR in a Trojan-horse approach where cytotoxic drugs, loaded onto CN, did not directly interact with the membrane machinery thus escaping the efflux transporters [131]. Like nanoparticle systems, CN nanocarriers accumulate in tumor masses through cell junction gaps (around 100–780 nm) of leaky vasculature with poor lymphatic drainage (EPR effect) [132]. Moreover, they are able to cross the cell membrane via dual mechanism involving endocytic pathway or passive diffusion [133]. In detail, individual CNTs are mainly internalized by passive diffusion due to their needle-like shape [134,135], whereas for clustered nanotubes, a clathrin-dependent endocytosis through endosomes followed by trafficking to lysosomes in the perinuclear compartment has been described [136,137] (Figure 3).

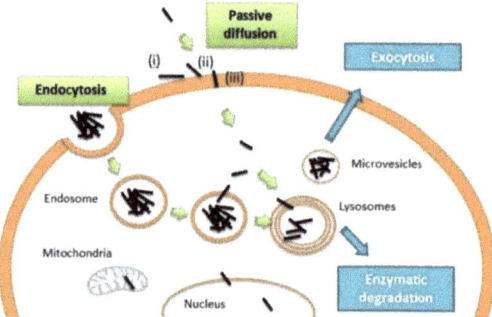

Figure 3. General representation of the main mechanisms involved in the CNT cellular uptake. Reproduced with permission from [137] Elsevier (2016).

Clathrin-dependent mechanism is also accepted as the main mechanism for intracellular internalization of GO and rGO [138]. C_{60} toxic side-effects due to cell membrane penetration have been reported, while endocytosis pathways, typical of hydrophilic C_{60} derivatives uptake, overcome this limitation [139]. The stealth effect can be finely tuned via derivatization of the CN surfaces with a wide range of chemical species, allowing the possibility to spatially control their biological activity by conjugation with targeting elements such as folic acid and antibodies [140,141].

3.2.1. Pristine and Non-Covalently Functionalized CN

Some key examples of MRD reversal by enhanced cellular uptake by pristine and non-covalently coated CN are reported in Table 3.

Table 3. MDR reversal by enhanced uptake of pristine and non-covalently functionalized CN.

Carrier			Delivery Properties			Cancer Model		Ref
CN	Derivatizing Agent	Bioactive Agent	DL	Responsivity	Tissue	In Vitro	In Vivo	
SWCNT	TCM π-π *Staking*	ETP*	11-88§	—	Pancreas	PANC-1	—	[142]
oxMWCNT	—	VER π-π *Stacking* DOXπ-π *Stacking*	149 164	—	Blood	K562/A02/R	—	[143]
SWCNT	—	CpG	—	—	Brain	K-Luc	—	[144]
						GL261		
					Ovary	OVCAR8		
					Cervix	HeLa		
GO	—	CDDP π-π *Staking*	400	—	Ovary	SCOV-3	—	[145]
					Cervix	HeLa		
					Prostate	Tramp-C1		
					Lung	A549		
					Colon	CT26		
GO	ASO *Hybridization*	DOX π-π *Staking*	35.25	—	Breast	MCF-7/ADR	MCF-7/ADR	[146]
usGWCNT	PF108 π-π *Staking*	CDDP *Filling*	6.4	—	Breast	MCF-7 MDA-MB-231	—	[147]
usSWCNT	PF108 π-π *Staking*	CDDP *Filling*	6.4	—	Breast	—	MCF-7 MDA-MB-231	[148]
oxMWCNT	—	Pt(IV) *Filling*	37	—	Cervix	HeLa	—	[149]

Table 3. Cont.

Carrier		Delivery Properties			Cancer Model			Ref
CN	Derivatizing Agent	Bioactive Agent	DL	Responsivity	Tissue	In Vitro	In Vivo	
GQD	—	CDDP π-π Staking	0-50	—	Liver	SMMC-7721	—	[150]
					Cervix	HeLa		
					Lung	A549		
					Breast	MCF-7		
					Stomach	MGC-803		
GQD	—	CDDP Condensation	—	pH	Os	HSC3	HSC3	[151]
	PEG-NH$_2$ Condensation					SCC4	—	
						CAL-27	—	
GQD	—	DOX π-π Staking	10	pH	Breast	MCF-7	—	[152]
						MCF-7/ADR		
					Stomach	MGC-803		
GO	—	DOX π-π Staking	47	pH	Breast	MCF-7	—	[153]
						MCF-7/ADR		
GO	HDex π-π Staking	DOX π-π Staking	350	pH	Breast	MCF-7/ADR	—	[154]
GO	HEC/PAC π-π Staking	DOX π-π Staking	49	pH	Ovary	SCOV-3	—	[155]
						SCOV-3/DDP		
SWCNT	DISPE-PEG π-π Staking	PTX/C6/QD π-π Stacking	14.3	Magnetic	Pancreas	PANC-1	—	[156]
						MIA PaCa-2		
						L3.6		
oxSWCNT	DISPE-HA π-π Staking	ERU π-π Stacking	45	pH	Lung	A549	—	[157]
						A549/TXR		
GO	PEI/PSS π-π Staking	DOX π-π Staking Anti-miR-21	—	—	Breast	MCF-7	—	[158]
						MCF-7/ADR		

* Co-administration with CN; § CNT/drug; DL: drug loading %; ASO: anti-sense oligonucleotide; C6: ceramide C6; CDDP: cisplatin; DISPE: distearoyl-sn-glycero-3-phosphoethanolamine; DOX: doxorubicin; ERU: epirubicin; ETP: etoposide; GO: graphene oxide; GQD: graphene quantum dots; HA: hyaluronic acid; HDex: hematin-dextran; HEC: hydroxyethyl cellulose; MWCNT: multi-walled carbon nanotubes; oxMWCNT: oxidized MWCNT; PAC: polyanionic cellulose; PEG: polyethylene glycol; PF: Pluronic F; PSS: poly(sodium 4-styrensulfonates); PTX: paclitaxel; QD: quantum dots; SWCNT: single-walled carbon nanotubes; OxSWCNT: oxidized SWCNT; TCM: tissue culture medium; UsSWCNT: ultra-short SWCNT; VER: verapamil.

The ability of pristine SWCNT to penetrate the cell membrane was exploited by Mahmood et al. [142] for the treatment of drug-resistant pancreatic cancer. The authors proved that SWCNT enhanced the cellular uptake of etoposide ETP by 2–5 times when co-administrated. The co-delivery of a cytotoxic agent (doxorubicin—DOX) and a P-gp inhibitor (verapamil—VER) by oxMWCNT, resulted in a significant improvement in the DOX anticancer efficiency due to the increased drug uptake by leukemia drug-resistant cells [143]. The hybridization of CpG oligonucleotide onto pristine SWCNT was used as a strategy to enhance the cell internalization and thus the activation of the innate immune system via Toll-like receptor 9 in a malignant brain cancer model [144]. As a consequence, a selective inhibition of glioma cells migration was observed, while macrophages viability and proliferation remained almost unaltered.

Unmodified GO either enhances the nuclear uptake of cisplatin (CDDP) in several cancer cell lines [145], or delivers oligonucleotides into cells protecting them from enzymatic cleavage [159]. The latter property was exploited by Li et al. [146], who developed a DOX carrier system based on GO modified with two molecular beacons (MBs). The intracellular delivery of MBs silenced the MDR1 and upstreamed erythroblastosis virus E26 oncogene homolog 1 mRNAs. This resulted in an effective inhibition of the P-gp expression and thus in an enhanced efficacy of DOX in resistant breast cancer cells.

The pyrolysis of Fluorinated SWCNT at 1000 °C in an argon atmosphere produced ultra-short single-walled carbon nanotubes (us-SWCNT), which resulted in negligible toxicity when administered in mice [160]. The inner cavity of usSWCNT was filled with CDDP and the resulting device proposed as delivery vehicle for the treatment of breast cancer both in vitro [147] and in vivo [148]. Pluronic 68 (PF 68) was used as coating element to reach an extension of the CDDP release profiles overtime, and an enhancement of drug cytotoxicity against MCF-7 and MDAMB-231 cells was observed in vitro as a consequence of the enhanced cellular uptake [147]. Furthermore, higher CDDP uptake in tumors was detected in in vivo experiments, due to the prolonged blood circulation time facilitating tumor targeting by the EPR effect.

The effect of the CNT diameter on the carrier efficiency against HeLa cells was investigated by Muzi et al. [149]. oxMWCNT with inner diameters of 10 and 38 nm were filled with a hydrophobic Platinum (IV) complex. The authors found that the larger CNT possessed higher cytotoxic properties, whilst the 10nm CNT provided a more prolonged payload release. Interestingly, both carriers were poorly cytotoxic on macrophages and did not induce any pro-inflammatory response.

The correlation between CNT size reduction and enhancement of the anticancer activity was also exploited in the case of GO materials. In detail, GQD were obtained by Fenton reactions of GO and used as nuclear uptake enhancers of CDDP [150,151] and DOX [152] in various solid cancers. pH-dependent vectorization of DOX to the nucleus of drug-resistant breast cancer cells was obtained by DOX loading onto GO nanosheets [153].

GO carriers enter cells via endocytosis and, escaping the drug efflux systems, allow an effective MDR reversal and a significant reduction of MCF-7/ADR viability. This was shown by the high reversal index value, expressed as the ratio between IC_{50} values of free DOX and DOX@GO. Similar results were obtained by a self-assembled G−dextran nanohybrid, fabricated by π−π interactions of GO and hematin-terminated dextran (HDex) [154], or a GO nanocarrier prepared by using hydroxyethyl cellulose (HEC) and polyanionic cellulose (PAC) as nonionic–anionic synergistic surfactants for GO stabilization in serum [155].

Distearoyl-sn-glycero-3-phosphoethanolamine (DISPE)-PEG was used as coating for the construction of SWCNT nanohybrids suitable for the vectorization of paclitaxel (PTX) and ceramide C6 into drug-resistant pancreatic cancer cells [156]. The key study result was the possibility to trigger the intracellular release of the payloads on-demand from the CNT inner core by inductive heating with an external alternating current or pulsed magnetic field. Negligible toxic side effects have been hypothesized due to the retention of drug inside the nanocarriers in the absence of the external stimulation. DISPE-HA coated SWCNT were used for the targeted delivery of epirubicin (EPI) to CD44-overexpressing resistant lung cancer cells [157]. In vitro experiments demonstrated that the system significantly increased the intracellular delivery and retention of EPI through CD44 receptor-mediated endocytosis.

A multifunctional nanocomplex, composed of GO, polyethylenimine (PEI) and poly(sodium 4-styrenesulfonates) (PSS) was used for the combined DOX delivery and miR-21 gene silencing in drug-resistant breast cancer cells [158]. miR-21 over-expression is significantly correlated with drug resistance in breast cancer, thus the simultaneous down-regulation of miR-21 gene and the enhanced cell accumulation of DOX was proposed as a valuable strategy for re-sensitizing resistant cells to the cytotoxic agent.

3.2.2. Covalently Functionalized CN

Kim et al. [161] investigated the effect of different CN conjugation types (e.g., covalent and non-covalent) on DOX localization in cancer cells. For example, DOX was either conjugated to oxMWCNT via amide bond, or absorbed onto PEG wrapped oxMWCNT via π-π stacking. The results showed a lower DOX uptake in normal cells than in cancer cells, while a higher cellular uptake by clathrin-dependent mechanisms was recorded in the case of covalent conjugation. Furthermore, the conjugation was more effective in sustaining DOX release inside cells, while a faster release was detected in the case of absorbed drug, due to the carrier vulnerability in both low acidic (pH < 5.0) and enzymatic environments. A specific intracellular fate was also observed: a non-covalent carrier was more advantageous for mitochondria drug delivery, while a nucleus targeted delivery was obtained with the DOX-conjugated MWCNT. Furthermore, the covalent conjugation resulted in less amount of effluxed DOX from cancer cells, greater apoptosis and cytotoxic activity. Similarly, the pro-apoptotic potential of ginseng secondary metabolites (ginsenoside Rb1 or Rg1) was improved upon conjugation to oxMWCNT [162].

Different covalent functionalization routes involve CN surface modification with polymeric materials and the subsequent loading of the bioactive agents via physical interaction or chemical conjugation to the hybrid carrier (Table 4).

Table 4. MDR reversal by enhanced cellular uptake of covalently functionalized CN.

Carrier		Delivery Properties			Cancer Model			Ref
CN	Derivatizing Agent	Bioactive Agent	DL	Responsivity	Tissue	In Vitro	In Vivo	
oxMWCNT	—	DOX Condensation	112	pH	Lung	A549	—	[161]
	PEG π-π Stacking	DOX π-π Stacking	31.4		Breast	MDA-MB-231		
oxMWCNT	RB1 Condensation	—	25	—	Breast	MCF-7	—	[162]
	RG1 Condensation	—			Pancreas	PANC-1		
oxMWCNT	PEG-NH$_2$ Condensation	RuPOP π-π Stacking	9.8	pH X-ray	Liver	HepG2	—	[163]
						R-HepG2		
oxSWCNT	PSE-PEG-NH$_2$ Condensation	TRAIL Condensation	61	—	Liver	HepG2	—	[164]
					Colon	HCT116		
					Lung	H1703		
GO	NH$_2$-PEG-N$_3$ Condensation	DOX π-π Stacking	78	pH	Lung	A549	A549	[165]
		TRAIL	8		Colon	LoVo	—	
GO	H$_2$N-PEG-PEI Condensation	Condensation CER Ionic SRB*	—	—	Liver	HepG2	—	[166]
						HuH7	HuH7	
							HuH7-SR	
						HepG2	—	
oxSWCNT	PEG-HBA/ PEG-CD44 Ab Condensation	PTX Condensation	180	pH	Breast	MDA-MB-231	MDA-MB-231	[167]
		SAL Condensation	170					
oxSWCNT	CS-FA Condensation	O$_2$ Complexation	—		Breast	MDA-MB-231	—	[168]
		5-FU*	3.3§					
		ERU*	20§			ZR-75-1		
		PRU*	125§					
		PTX*	21§					
		CBPT*	10§					
GQD	HA-PEG-NH$_2$ Condensation	DOX π-π Stacking	30	pH	Lung	A549	—	[169]
GQD	HA-HSA NPs Condensation	GEM π-π Stacking	16	—	Pancreas	Panc-1	—	[170]
rGO	DEX-CT Redox coupling	DOX π-π Stacking	20	pH	Neural Crest	BE(2)C	—	[171]
						BE(2)C/ ADR		
oxSWCNT	P-gp Ab Condensation	DOX π-π Stacking	20	NIR	Blood	K562	—	[172]
						K562R		

Table 4. Cont.

Carrier		Delivery Properties			Cancer Model			Ref
CN	Derivatizing Agent	Bioactive Agent	DL	Responsivity	Tissue	In Vitro	In Vivo	
oxMWCNT	P-gp Ab Condensation	DOX π-π Stacking GA π-π Stacking	39.4 30.3	pH	Blood	K562/A02/R	K562/A02/R	[173]
oxSWCNT	CD133 Ab Condensation	CDDP π-π Stacking/ Condensation	66	—	Skin	B16-F10	B16-F10	[174]
		Pt(IV) π-π Stacking/ Condensation	66					
GO	FA-PAMAM-DTPA Condensation	DOXnπ-π Stacking COLCnπ-π Stacking	154 154	pH	Liver	HepG2	HepG2	[175]
GO	PAMAM Condensation	DOX π-π Stacking sRNA Hybridization	28.6 5	pH	Breast	MCF-7	—	[176]
GO	FA-CO Condensation	DOX π-π Stacking sRNA Hybridization	56	pH	Breast	MCF-7	—	[177]
						MCF-7/ADR		
					Lung	A549		
C_{60}	Br-C-(COOEt)$_2$ Bingel	CDDP*			Prostate	PC-3	PC-3R	[178]

* Co-administration with CN; § CNT/drug; DL: drug loading % (w/w); 5-FU: 5-fluorouracil; Ab: antibody; CBPT: carboplatin; CDDP: cisplatin; CER: ceramide; CO: chitosan oligosaccharide; COLC: colchicine; CS: chitosan; CT: catechin; DEX: dextran; DOX: doxorubicin; DTPA: diethylenetriamine pentaacetate; ERU: epirubicin; FA: folic acid; GA: gambogic acid; GEM: gemcitabine; GO: graphene oxide; HA: hyaluronic acid; HBA: 4-hydrazinobenzoic acid; HSA: human serum albumin; MWCNT: multi-walled carbon nanotubes; oxMWCNT: oxidized MWCNT; NPs: nanoparticles; PAMAM: polyamidoamine; PEG: polyethylene glycol; PEI: polyethylenimine; POP: polypyridyl; PRU: pirarubicin; PSE: 1-pyrenebutanoic acid; PTX: paclitaxel; RB1: RB1 ginsenoside; RG1: RG1 ginsenoside; rGO: reduced graphene oxide; SAL: salinomycin; SRB: sorafenib; SWCNT: single-walled carbon nanotubes; oxSWCNT: oxidized SWCNT; TRAIL: tumor necrosis factor-related apoptosis-inducing ligand.

For example, CN covalent functionalization with PEG enhanced the cell internalization of ruthenium polypyridyl complex (RuPOP) [163] and Tumor necrosis factor-related apoptosis-inducing ligand (TRAIL) [164,165] in different cancer cell lines. In the first case, the enhanced cell internalization of PEG-SWCNT resulted in a significant radio-sensitizing effect and in the possibility to kill resistant liver cancer cells by X-ray treatment, while TRAIL efficacy was enhanced by either PEGylated CNT [164] or GO [165]. In the latter case, a pH-responsive co-delivery of DOX to lung and colon cancers was also demonstrated, both in vitro and in vivo [165].

The covalent derivatization of GO with PEG-polyethylenimine (PEI) was proposed for the co-delivery of sorafenib (SOR) and CER for the treatment of drug-resistant liver cancer [166]. PEG-functionalized SWCNT were derivatized via hydrazone linkers with both salinomycin or PTX as anticancer agents and CD44 antibody for the treatment of breast cancer and cancer stem cells (CSC) subpopulation [167]. pH-responsive release mechanism near the acidic tumor microenvironment was observed, and the in vivo therapeutic efficacy was shown in tumor bearing mice, confirming the therapeutic efficacy of the proposed formulation.

By covalent derivatization of oxSWCNT with chitosan-folic acid conjugate (CS-FA), Jia et al. [168] developed a carrier targeting the hypoxic environment of breast cancer. Interestingly, since the adaptation processes developed by cancer cells to proliferate in the hypoxic environment are also at the basis of MDR insurgence [179], oxygen supply by CNT was found to be an effective approach for MDR reversal by counteracting the imbalance between oxygen demand and supply of cancer tissues [180].

The HA ability to target cancer cells was the rationale of Luo's [169] and Nigam's [170] research groups, who exploited the fast and high cell internalization rate of GQD for the vectorization of DOX and gemcitabine (GEM) to drug-resistant lung and pancreas cancer cells, respectively. In the first case, PEGylated GQD was proposed as pH-responsive vehicle for the intracellular delivery of DOX, while the conjugation with human serum albumin was found to enhance the tumor infiltration via gp60 pathway for overcoming the inadequate cellular uptake and small half-life of GEM.

rGO nanohybrids obtained by reductive coupling with an enzymatically synthesized dextran-catechin (DEX-CT) conjugate [181] were proposed as pH responsive delivery vehicle of DOX to either drug sensitive (BE(2)C) or drug-resistant (BE(2)C/ADR) neuroblastoma cells [171]. Taking advantage from the ability of CT moieties to down-regulate P-pg expression, authors proved the possibility to synergize the DOX activity in BE(2)C cells and promote resistance reversal in BE(2)C/ADR cells.

Active targeting strategies were proposed by Li et al. [172]. In this study, anti-P-gp SWNTs functionalized with anti-P-gp antibody via amide bonding were used as DOX vehicle for the treatment of leukemia. By comparing the efficiency of the nanocarriers on drug-resistant K562R and drug sensitive K562S cells, a 23-fold higher binding affinity and a specific localization on the cell membrane of K562R cell were recorded, with a 2.4-fold higher cytotoxic activity. Zhang et al. [173] proposed a further upgrade of this concept by co-loading DOX and gambogic acid (GA) as a cytotoxic agent and a P-gp inhibitor, respectively. As a result, high accumulation of anticancer drug in leukemia drug-resistant cells, and a relevant cytotoxic effect due to enhanced apoptosis were recorded both in vitro and in vivo.

A similar approach was used by Nowacki et al. [174], where anti-CD133 antibodies and CDDP complexes were employed as targeting and bioactive agents, respectively. The authors investigated the effect of either physical or chemical loading on peritoneal carcinomatosis.

Zhang et al. [175] reported the covalent functionalization of GO with a gadolinium-labeled polyamidoamine (PAMAM) dendrimer for in vivo imaging and liver cancer targeted therapy based on the synergistic combination of DOX and colchicine (COLC). PAMAM generation 3.0 functionalized GO nanosheets were also proposed for the pH-responsive delivery of DOX in the presence of MMP-9 shRNA in breast cancer cells, reaching a transfection efficiency significantly higher than that obtained with conventional polymeric carriers such as PEI [176]. A similar approach was developed by Cao et al. [177], who employed a FA-chitosan oligosaccharide (CO) conjugated as derivatizing agent. According to this approach, FA acts as targeting unit, while the amino groups of CO are responsible for the delivery of MDR1 siRNA, allowing an effective MDR reversal in breast and lung cancer cells and an improved DOX efficiency. Finally, it should be cited the covalent derivatization of C_{60} through the Bingel reaction as a strategy for the enhanced cellular uptake of CDDP to either drug sensitive or resistant prostate cancer cells [178].

3.3. MDR Reversal by Cell Damage

The interaction between nanomaterials and cell environment is related to both their morphological features and surface properties [182].

Fiber-like materials such as CNT where found to possess intrinsic capability to inhibit microtubule dynamics during mitosis, thus reducing the cell replication rate [183]. Furthermore, toxicological studies showed that, upon exposure to CN, an imbalance between the production of reactive oxygen species and their detoxification by biological systems occurs [184]. This phenomenon ignites a cascade of biological responses, including destabilization of cells and mitochondrial membranes and eventually induction of apoptosis [185], mainly through the MAPK and TGF-β signaling pathways [186]. Such intrinsic cytotoxicity can be exploited for contrasting MDR acquisition in cancer cells, due to their fast replication rate and their high sensitivity to oxidative stress [187] (Table 5).

Table 5. MDR reversal by CN via cell damage.

Carrier		Delivery Properties			Cancer Model			Ref
CN	Derivatizing Agent	Bioactive Agent	DL	Responsivity	Tissue	In Vitro	In Vivo	
MWCNT	TCM π-π Stacking	—	—	—	Skin	B16-F10	B16-F10	[188]
MWCNT	5-FU π-π Stacking		3	—	Skin	—	B16-F10	[189]
					Cervix	HeLa	—	
GO	NH_3 Oxidation	—	—	pH	Cervix	HeLa	HeLa	[190]
rGO	Ag NPs Redox Coupling	TSA π-π Stacking	100	—	Ovary	SKOV-3	—	[191]
GO	PEG Condensation FePt MNPs π-π Stacking	MI π-π Stacking 5-FU π-π Stacking	12.3 9.5	O_2	Lung	A549 H1975	—	[192]

DL: drug loading % (w/w); 5-FU: 5-fluorouracil; GO: graphene oxide; MI: metronidazole; MNPs: magnetic nanoparticles; MWCNT: multi-walled carbon nanotubes; NPs: nanoparticles; PEG: polyethylene glycol; rGO: reduced graphene oxide; TCM: tissue culture medium; TSA: trichostatin A.

García-Hevia et al. [188] showed that MWCNT can form biosynthetic microtubules with tubulin responsiveness, resulting in severe deficiencies during mitosis, inhibition of cell migrations, and cell death. Interestingly, the same effect was observed in a xenograft model of MDR melanoma, demonstrating the need for further investigations. In another work from the same group [189], 5-Fluoruracil (5-FU) was used as both surface derivatizing element and bioactive agent. The aim was to couple the CNT key features and drug counterpart in a single device, obtaining a synergistic effect between the ability of 5-FU to inhibits cell replication in the "S" phase and the effect of MWCNT on microtubule dynamics.

Lin et al. [190] functionalized GO by chemical oxidation in the presence of NH_3 and H_2O_2 (N-GOs), obtaining a nanomaterial with peroxidase-like activity, able to disproportionate H_2O_2 into hydroxyl radicals in the acidic microenvironment of tumor cells, triggering cell necrosis in vitro and in vivo.

A different approach for the modulation of ROS levels in drug-resistant cancer cells involves the functionalization of GO with metal nanoparticles such as Ag [191] and FePt [192] MNPs. In the first case, the authors synthesized rGO-Ag nanosystem using lycopene as reducing agent, and provided scientific evidence that rGO-Ag promotes ROS generation sensitizing human ovarian cancer cells to trichostatinA (TSA) and inducing cell death. In the second study, Fe-Pt was adsorbed onto PEGylated GO to obtain ROS overproduction and synergize metronidazole (MI) and 5-FU toxicity for the treatment of lung cancer cells.

3.4. MDR Reversal by Phototherapy

Therapies relying on light are emerging as valuable tools for fighting cancer, by taking advantage of the site-specificity, the poor insurgence of side effects, the absence of cell resistance, as well as the possibility to trigger the release of cytotoxic drugs to the disease site [193].

In this regard, phototherapies can be classified in two main categories, photothermal and photodynamic therapy, differing from approach and working mechanism. In the first case, nanomaterials are able to absorb light and produce heat, with the consequent thermal ablation of cancer cells [194]. Photodynamic protocols are based on the generation of singlet oxygen and other ROS, upon exposure to light [195]. In both cases, wavelength in the near-infrared (NIR) region are used since they can reach deeper sites in the body. CNT, C_{60} and GO exhibit high NIR-absorbing capability, and thus are widely proposed as nanocarriers for combined chemo- and photo- therapies (Figure 4) [196,197].

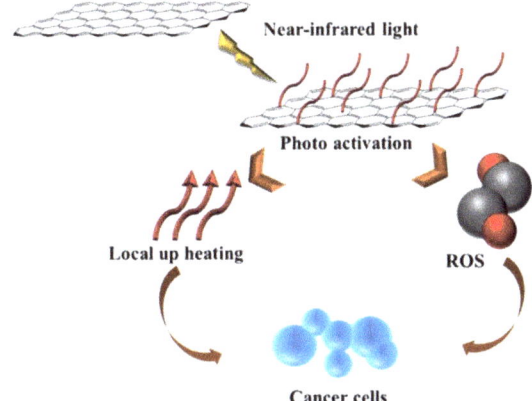

Figure 4. General representation of photothermal therapy by G derivatives. Adapted with permission from [198] copyright Elsevier (2018).

3.4.1. Pristine and Non-Covalently Coated CN

Table 6 summarizes the recent examples of MRD reversal by photo-thermal ablation obtained by the employment of un-modified or non-covalently functionalized CN. C_{60} nanocrystals (nC_{60}) are interesting CN obtained upon contact of un-modified C_{60} with water. Under specific conditions, C_{60} forms a water-stable, colloidal aggregate with reported diameters in the 5–500nm range [199]. nC_{60} were found to be suitable materials for the photo-dynamic therapy of thermal ablation of cervix and breast cancer cells [200]. Interestingly, the ROS production and ability to eliminate cancer cells have been shown to significantly increase by wrapping a Neodymium atom in the C_{60} spherical cage [201].

Table 6. MDR reversal by photo-thermal ablation induced by pristine and non-covalently functionalized CN.

Carrier			Delivery Properties			Cancer Model			Ref
CN	Derivatizing Agent	Bioactive Agent	DL	Responsivity	Tissue	In Vitro	In Vivo		
nC_{60}	—	DOX*	—	—	Cervix	HeLa	—		[200]
					Breast	MCF-7/ADR			
nC_{60}	Nd Encapsulation	—	—	—	Cervix	HeLa	—		[201]
						H1975			
MWCNT	PEG π-π Stacking	—	—	NIR	Pancreas	PANC1	—		[202]
oxMWCNT	$H_2NC_2H_4NH_2$ Amidation DISPE-PEG π-π Stacking	PTX*	—	NIR	Breast	HMLER	—		[203]
		SAL*				HMLER CSC	HMLER CSC		
		17DMAG*							
oxSWCNT	DISPE-PEG/P-gp Ab π-π Stacking	—	—	—	Fibroblast	3T3-MDR1	—		[204]
					Ovary	NCI/ADR			
oxCNH	PEG/P-gp Ab π-π Stacking	ETP Filling	—	NIR	Lung	A549			[205]
						A549R	A549R		
SWCNT	CA-HA π-π Stacking	DOX π-π Stacking	300	—	Ovary	OVCAR8	—		[206]
						OVCAR8/ADR	OVCAR8/ADR		
oxSWCNT	CS-FA π-π Stacking	DOX π-π Stacking	33.3	NIR	Lung	A549	A549		[207]
SWCNT	CS-CD133 Ab π-π Stacking	—	—	NIR	Brain	GMB-CD133+	GMB-CD133+		[208]
						GMB-CD133-	GMB-CD133-		

Table 6. *Cont.*

Carrier			Delivery Properties			Cancer Model		Ref
CN	Derivatizing Agent	Bioactive Agent	DL	Responsivity	Tissue	In Vitro	In Vivo	
SWCNT	GCS π-π Stacking	—	—	NIR	Breast	EMT6	EMT6	[209]
GO	PEGylated Liposome Encapsulation	DOX π-π Stacking RAPA π-π Stacking	10 10	pH	Breast	MCF-7 MDA-MB-231 BT4T4	—	[210]
GO	PF 68	DOX π-π Stacking IRI π-π Stacking	7 7	pH	Breast	MCF-7 MDA-MB-231	—	[211]
					Head/Neck	SCC-7		

* Co-administration with CN; DL: drug loading % (*w/w*); 17DMAG: 17-(dimethylaminoethylamino)-17-demethoxygeldanamycin; Ab: antibody; nC$_{60}$: C$_{60}$ nanocrystals; CA cholanic acid; CNH: carbon nanohorns; CS: chitosan; DISPE: distearoyl-sn-glycero-3-phosphoethanolamine; DOX: doxorubicin; ETP: etoposide; FA: folic acid; GCS: glycated chitosan; GO: graphene oxide; HA: hyaluronic acid; IRI: irinotecan; MWCNT: multi-walled carbon nanotubes; NIR: near infrared radiation; oxCNH: oxidized CNH; oxMWCNT: oxidized MWCNT; oxSWCNT: oxidized SWCNT; PEG: polyethylene glycol; PF: Pluronic F; PTX: paclitaxel; RAPA: rapamycin; SAL: salinomycin; SWCNT: single-walled carbon nanotubes.

The NIR-absorbing properties of MWNCT in pancreas cancer cells were investigated by Mocan et al. [202], proving that, upon laser treatment, PEGylated MWCNT nanohybrids induced immediate cellular apoptosis as a consequence of increased mitochondrial membrane depolarization.

DISPE-PEG coated CNT were proposed as tools for photo- chemo- therapy protocols involving NIR irradiation of nanohybrid in combination with conventional cytotoxic drugs with the aim to find an effective therapeutic approach for the treatment of chemo- and radio-resistant breast cancer stem cells [203]. An upgrade of this concept was obtained by introducing P-gp antibodies as derivatizing agents for MWCNT surface [204], reaching a targeted thermal ablation in tumor spheroids of MDR cancer cells, with absence of side toxicity on healthy cells. Similarly, P-gp antibodies were conjugated on the surface of oxidized CNH for a combined chemo- and photo-thermal therapy of non-small cell lung cancer [205].

The strategy proposed by Bhirde and co-workers [206] is based on the employment of cholanic acid-derivatized hyaluronic acid (CA-HA) as targeting element in a SWCNT-DOX based therapy for ovarian cancer. Authors stated that the employed noncovalent approach, by preserving the surface integrity and properties of SWCNT, was able to vectorize the drug (due to HA moieties) and improve the cytotoxic drug uptake into either sensitive or resistant cancer cells. A significant reduction of the resistance factor (expressed as the ratio between the IC$_{50}$ values of drug-resistant to drug-sensitive cells) was obtained, and the combination with NIR irradiation was employed for a further enhancement of the therapeutic efficiency in vivo.

Selective vectorization of DOX was achieved by using CS-FA as coating material for oxSWCNT, leading to a NIR-induced intracellular delivery of DOX, with a 12-fold decrease of the IC$_{50}$ of DOX in lung cancer cells [207]. Wang et al. [208] produced anti-CD133 functionalized SWNTs for the selective thermal ablation of glioblastoma stem-like cells in vivo.

The employment of coating with intrinsic biological activity was the rationale of the study by Zhou et al. [209]. By coating SWCNT with glycated chitosan (GCS), the authors developed an immune-adjuvant nanohybrid able to enhance the tumor immunogenicity, leading to remarkable antitumor activity in vivo due to the combined thermal and immunological effects. Nanocarriers, obtained by GO coating with surfactant materials, were developed for the combined photo-thermal treatment of breast and head and neck cancers. PEGylated lipid bilayers [210] and PF 68 [211] were proposed as GO wrapping agents for the synthesis of pH responsive nanocarriers in order to deliver DOX in combination with rapamycin (RAPA) or Irinotecan (IRI), respectively. Upon NIR irradiation, significant apoptosis was induced in breast [210,211] and head and neck [211] cancer cells, resulting in an effective chemo/photo-thermal therapy due to the simultaneous thermal ablation.

3.4.2. Covalently Functionalized CN

Covalent functionalization strategies were also proposed for the development of CN carrier systems with the goal to co-adjuvate cancer therapy by photo-thermal MDR reversal (Table 7).

Table 7. MDR reversal by photo-thermal ablation induced by covalently functionalized CN.

Carrier		Delivery Properties			Cancer Model			Ref
CN	Derivatizing Agent	Bioactive Agent	DL	Responsivity	Tissue	In Vitro	In Vivo	
GO	P-gp Ab Condensation FA-Au NPs π-π Stacking	MiR-122 Hybridization	—	—	Liver	Hep-G2/ADR	Hep-G2/ADR	[212]
GO	HA Condensation PF 68 π-π Stacking	MIT π-π Stacking	3	pH	Breast	MCF-7 MCF-7/ADR	MCF-7 MCF-7/ADR	[213]
GO	PF 68-PAMAM Diselenide	ICG π-π Stacking	52.1	ROS	Breast	MCF-7 MCF-7/ADR	—	[214]
GO	PF 68-PAMAM Diselenide	ICG π-π Stacking	52.1	ROS	Breast	MCF-7 MCF-7/ADR	— MCF-7/ADR	[215]
GO	PEG-PAH Condensation	DOX π-π Stacking	50	pH	Breast	MCF-7 MCF-7/ADR	—	[216]
GO	FA-PEG-PEI Condensation	DOX π-π Stacking sRNA Hybridization	—	pH	Breast	MCF-7 MCF-7/ADR	—	[217]
GO	PEG-NH$_2$ Condensation HPPH π-π Stacking	CTX π-π Stacking DOX π-π Stacking DTX π-π Stacking 5-FU π-π Stacking	1	NIR	Breast	4T1	4T1	[218]
GO	TRF Condensation	—	—	—	Blood	K562 K562R	—	[219]
GO	Fe$_3$O$_4$/MnO$_x$ Redox Coupling	DOX π-π Stacking	38	pH Redox Magnetic	Breast	MDA-MB-231 MCF-7/ADR	—	[220]

5-FU: fluorouracil; Ab: antibody; CTX: cyclophosphamide; DOX: doxorubicin; DTX: docetaxel; FA: folic acid; GO: graphene oxide; HA: hyaluronic acid, HPPH: 2-(1-hexyloxyethyl)-2-devinyl pyropheophorbide a; ICG: indocyanine green; MIT: mitoxantrone; NPs: nanoparticles; PAH: poly(allylamine hydrochloride); PAMAM: polyamidoamine; PEG: polyethylene glycol; PEI: polyethylenimine; PF: Pluronic F; TRF: transferrin.

An improved photo-thermal protocol was developed by Yuan et al. [212]. Here, GO-Gold nanoparticles functionalized with FA and anti- P-gp antibody were used as carrier for MiR-122, allowing an effective induction of apoptosis in drug-resistant HepG2 liver cancer cells, with the possibility to combine drug targeting and controlled release.

The functionalization of GO with HA units conferred targeting ability as well as the possibility to modulate the release of mitoxantrone (MIT) by NIR irradiation [213]. The resulting nanocarrier was also able to act as reversible inhibitor of P-gp, enhancing drug efficiency in either drug-sensitive or drug-resistant breast cancer cells, in vitro and in vivo. A further improvement of PF 68- based GO nanohybrid was proposed by the Wang research group, by developing ROS responsive nanocarriers for the combined vectorization of indocyanine green (ICG) and DOX as photosensitizer and chemo-therapeutic drug, respectively, in breast cancer cells [214,215]. The synthetic approach is based on the formation of a ROS sensitive diselenide bond between GO and a PF68-PAMAM conjugate acting as coating element. Each component contributes to the carrier efficiency: GO enhances the affinity to both photosensitizer and cytotoxic agents, PF 68 confers high stability in physiological environment to the whole system, PAMAM acts as proton sponge allowing lysosomal escape, while the NIR-absorbing properties of both GO and ICG trigger the drug release inside cells by the cleavage of the diselenide bond and eliminate cancer cell by thermal ablation. The covalent modification of GO with PEG in combination with a second polymer, such as poly(allylamine hydrochloride) (PAH), resulted in a pH-responsive DOX treatment of resistant breast cancer cells [216].

Zeng et al. [217] proposed the combination of PEG and PEI as GO derivatizing agent to generate a carrier system suitable for the vectorization of DOX and P-gp siRNA to resistant breast cancer cells. The enhanced carrier efficacy was due to the combination of increased loaded DOX, the modulation of P-gp expression by siRNA and the simultaneous thermal ablation upon laser treatment. Zhao et al. [218] used a photosensitizer (2-(1-hexyloxyethyl)-2-devinyl pyropheophorbide-a – HPPH) in combination with PEG for manufacturing hybrid GO nanocarriers for the photo-dynamic therapy of breast cancer in vitro and in vivo. Interestingly, the simultaneous treatment with conventional cytotoxic drugs induced an increase in tumor macrophage infiltration, resulting in an effective cancer eradication.

Alternative strategies for inducing thermal ablation of cancer cells involved the use of radiofrequencies and magnetic fields as activating agents. Sasidharan et al. [219] developed a targeted GO nanosystem by exploiting the overexpression of transferrin (TRF) on drug-resistant cancer cells to trigger thermal ablation of leukemia cells with almost no side-toxicity. Finally, the GO derivatization with superparamagnetic Fe_3O_4 NPs carried out for the preparation of DOX nanocarriers able to respond to an external magnetic field, including magnetic hyperthermia, in a combination therapy protocol for drug-resistant breast cancer cells [220].

4. Conclusions and Perspectives

MDR represents the main obstacle for the success of chemo-therapeutic protocols in cancer treatment, and great efforts are devoted to the investigation of new strategies for overcoming this phenomenon. Among others, nanoparticle systems such CN, offer solutions by virtue of their ability to interfere with cell structures and functions responsible for intrinsic and acquired MDR. Two main mechanisms are involved, namely the direct modulation of cell pathways and the effective intracellular delivery of MDR reversing agents (e.g., efflux pumps inhibitor or modulators of redox cell state).

Among the different CN, CNT and GO have attracted a higher interest, with promising results in in vitro and in vivo models of different cancers. Furthermore, the availability of several functionalization routes allows the surface properties of CN to be finely modulated by selecting the proper derivatizing agents able to address the specific therapeutic needs.

An overview of the obtained results is given in Table 8, where the outcomes, strength and weakness of each functionalization route is showed in terms of proved success (%) in three categories, namely direct MRD reversal, enhancement of the efficiency of a conventional cytotoxic drug, and the reduction of undesired side effects. Furthermore, for each group, the number of studies (%) covering a specific cancer model and the employed anticancer drug is also reported, in order to give an exhaustive overview of the state of the art in the field.

Table 8. Outcomes, strength and weakness of CN based system proposed for MDR reversal reviewed in this paper. Percentages are calculated based on the total studies, with some papers covering more studies simultaneously.

CNs	Deriv	Ref	Total Studies	Drug	Cancer Model	Direct MDR Reversal	Enhanced Drug Efficiency	Reduced Side Effects
				Studies (%)		Success (%)		
F	—	[178] [200] [201]	4	None (25) DOX (50) CDDP (25)	Cervix (50) Prostate (25) Breast (25)	75 * 25 #	75 * 25 #	0 * 0 #
CNT	—	[124] [128] [142] [144] [147] [148] [188] [189]	11	None (46) 5-FU (18) CDDP (18) TAM (9) ETP (9)	Breast (28) Cervix (18) Skin (18) Brain (9) Ovary (9) Colon (9) Pancreas (9)	55 * 18 #	36 * 18 #	0 * 9 #
CNT	Ox	[130] [143] [149] [161] [162] [164] [172] [173] [174]	16	None (6) DOX (30) TRAIL (19) RB1 (13) RG1 (13) Pt(IV) (13) CDDP (6)	Breast (18.5) Blood (18.5) Lung (13) Pancreas (13) Skin (13) Cervix (6) Bone (6) Liver (6) Colon (6)	50 * 25 #	94 * 19 #	31 * 0 #
GO	—	[129] [145] [146] [150] [151] [152] [153] [190] [191] [212] [220]	23	None (22) CDDP (48) DOX (26) TSA (4)	Breast (26) Cervix (13) Liver (13) Ovary (9) Lung (9) Stomach (9) Colon (9) Blood (4) Prostate (4) Os (4)	12 * 4 #	18 * 2 #	10 * 1 #
CNT	PEG	[127] [161] [163] [167] [202]	9	None (45) DOX (22) PTX (11) SAL (11) RuPOP (11)	Breast (34) Liver (23) Cervix (11) Blood (11) Pancreas (11) Lung (11)	56 * 0 #	56 * 22 #	33 * 11 #
CNH	PEG	[205]	1	ETP (100)	Lung (100)	100 * 100 #	100 * 100 #	0 * 0 #
GO	PEG	[151] [165] [169] [170] [192] [210] [216] [218]	12	DOX (50) 5-FU (17) CDDP (8) GEM (8) CTX (8) DTX (8)	Breast (50) Lung (25) Os (8) Colon (8) Pancreas (8)	75 * 42 #	92 * 42 #	42 * 0 #
CNT	Surf	[147] [148] [156] [157] [203] [204]	9	None (22) CDDP (22) PTX (22) SAL (11) 17DMAG (11) ERU (11)	Breast (56) Fibroblast (11) Ovary (11) Pancreas (11) Lung (11)	78 * 33 #	67 * 44 #	56 * 0 #
GO	Surf	[211] [213]	3	DOX (67) MIT (33)	Breast (67) Head and Neck (33)	33 * 33 #	100 * 33 #	0 * 0 #
GO	PEI	[158] [166] [217]	3	DOX (67) SRB (33)	Breast (67) Liver (33)	100 * 33 #	100 * 33 #	0 * 0 #
GO	Dend	[175] [176] [214] [215]	4	DOX (50) ICG (50)	Breast (75) Liver (25)	100 * 50 #	100 * 50 #	25 * 25 #

Table 8. Cont.

CNs	Deriv	Ref	Total Studies	Drug	Cancer Model	Direct MDR Reversal	Enhanced Drug Efficiency	Reduced Side Effects
				Studies (%)		Success (%)		
CNT	PS	[168] [206] [207] [208] [209]	9	None (22) DOX (22) 5-FU (11) ERU (11) PRU (11) PTX (11) CBPT (11)	Breast (67) Brain (11) Ovary (11) Lung (11)	33 * 33 #	78 * 22 #	11 * 0 #
GO	PS	[154] [155] [171] [177]	5	DOX (100)	Breast (40) Ovary (20) Lung (20) Neural Crest (20)	80 * 0 #	100 * 0 #	20 * 0 #
GO	PR	[219]	1	None (100)	Blood (100)	100 * 100 #	0 * 0 #	100 * 0 #

* In vitro; # In vivo; CN: carbon nanostructure; Deriv: derivatization; 5-FU: 5-fluorouracil; 17DMAG: 17-(dimethylaminoethylamino)-17-demethoxygeldanamycin; CBPT: carboplatin; CDDP: cisplatin; CNT: carbon nanotubes; CTX: cyclophosphamide; DOX: doxorubicin; DTX: docetaxel; ERU: epirubicin; ETP: etoposide; F: fullerenes; GEM: gemcitabine; GO: graphene oxide; ICG: indocyanine green; MIT: mitoxantrone; Ox: oxidation; PEG: polyethylene glycol; POP: polypyridyl; PTX: paclitaxel; PR: protein; PRU: pirarubicin; PS: polysaccharides; RB1: RB1 ginsenoside; RG1: RG1 ginsenoside; SAL: salinomycin; SRB: sorafenib; Surf: surfactant; TAM: tamoxifen; TRAIL: tumor necrosis factor-related apoptosis-inducing ligand; TSA: trichostatinA.

Although the reviewed studies are very heterogeneous and there is not a biological parameter to be unequivocally used for a direct comparison between the obtained results, some important considerations can be done by considering the score of each item reported in the table.

Most studies investigated the efficiency of carrier systems based on oxCNT, GO, and PEGylated CNT/GO for the treatment of different solid (mainly breast and cervix) and blood cancers. Furthermore, a relevant amount of studies reported the ability of CN materials to directly reverse the MDR allowing, at the same time, the enhancement of the efficiency of a co-administered drug (mainly DOX) in vitro. These results, being obtained via different molecular mechanisms, can be considered of great interest for scientists working in the field, because of the availability of different strategies, each effecting a peculiar biological pathway. However, despite the high performances recorded in some cases (e.g., carriers based PEGylated CNT or GO functionalized with dendrimers), several issues need to be overcome before hypothesizing a translation of CN nanosystems into clinical practice. At first, more extensive in vivo studies need to clarify the real extent of the obtained results, as highlighted by the differences between in vitro and in vivo success score in the direct MDR reversal and enhanced drug efficiency items of Table 8. Subsequently, the concerns about long-term toxicity must be considered, especially in the case of CNT-based devices, to address the great debate about the benefit of CN in the clinic. Also in this case, the in vivo score of reduced toxic side effects item of Table 8 is very low, even if more encouraging results seems to be achieved using GO as core element of multifunctional vehicles, for which a more homogeneous score is recorded between the effectiveness in MDR reversal, enhancement of drug efficiency, and the reduction of toxicology profiles. To this regard, the introduction of GDQ, coupling high biocompatibility and enhanced cell penetrating behavior, is a step forward for the development of even more interesting materials, although only few examples are available in literature.

Overall, more extensive preclinical and clinical studies are required, in a dynamic interdisciplinary exchange of knowledge between chemists, materials scientists, biologists, and oncologists. Only combining different and complementary expertise, we can hope to succeed in facing the challenges of MRD reversal.

Funding: This research received no external funding.

Conflicts of Interest: The authors declare no conflict of interest.

References

1. Frank, D.; Tyagi, C.; Tomar, L.; Choonara, Y.E.; du Toit, L.C.; Kumar, P.; Penny, C.; Pillay, V. Overview of the role of nanotechnological innovations in the detection and treatment of solid tumors. *Int. J. Nanomed.* **2014**, *9*, 589–613. [CrossRef]
2. Iyer, A.K.; Singh, A.; Ganta, S.; Amiji, M.M. Role of integrated cancer nanomedicine in overcoming drug resistance. *Adv. Drug Deliver. Rev.* **2013**, *65*, 1784–1802. [CrossRef] [PubMed]
3. Tsuruo, T.; Naito, M.; Tomida, A.; Fujita, N.; Mashima, T.; Sakamoto, H.; Haga, N. Molecular targeting therapy of cancer: Drug resistance, apoptosis and survival signal. *Cancer Sci.* **2003**, *94*, 15–21. [CrossRef] [PubMed]
4. Singh, M.S.; Tammam, S.N.; Boushehri, M.A.S.; Lamprecht, A. MDR in cancer: Addressing the underlying cellular alterations with the use of nanocarriers. *Pharmacol. Res.* **2017**, *126*, 2–30. [CrossRef]
5. Fojo, T.; Bates, S. Strategies for reversing drug resistance. *Oncogene* **2003**, *22*, 7512–7523. [CrossRef]
6. Gottesman, M.M.; Fojo, T.; Bates, S.E. Multidrug resistance in cancer: Role of ATP-dependent transporters. *Nat. Rev. Cancer* **2002**, *2*, 48–58. [CrossRef]
7. Peetla, C.; Vijayaraghavalu, S.; Labhasetwar, V. Biophysics of cell membrane lipids in cancer drug resistance: Implications for drug transport and drug delivery with nanoparticles. *Adv. Drug Deliver. Rev.* **2013**, *65*, 1686–1698. [CrossRef]
8. Housman, G.; Byler, S.; Heerboth, S.; Lapinska, K.; Longacre, M.; Snyder, N.; Sarkar, S. Drug Resistance in Cancer: An Overview. *Cancers* **2014**, *6*, 1769–1792. [CrossRef]
9. Kunz-Schughart, L.A.; Dubrovska, A.; Peitzsch, C.; Ewe, A.; Aigner, A.; Schellenburg, S.; Muders, M.H.; Hampel, S.; Cirillo, G.; Iemma, F.; et al. Nanoparticles for radiooncology: Mission, vision, challenges. *Biomaterials* **2017**, *120*, 155–184. [CrossRef]
10. Bakry, R.; Vallant, R.M.; Najam-Ul-Haq, M.; Rainer, M.; Szabo, Z.; Huck, C.W.; Bonn, G.K. Medicinal applications of fullerenes. *Int. J. Nanomed.* **2007**, *2*, 639–649.
11. Tasis, D.; Tagmatarchis, N.; Bianco, A.; Prato, M. Chemistry of carbon nanotubes. *Chem. Rev.* **2006**, *106*, 1105–1136. [CrossRef] [PubMed]
12. Wang, Y.; Li, Z.H.; Wang, J.; Li, J.H.; Lin, Y.H. Graphene and graphene oxide: Biofunctionalization and applications in biotechnology. *Trends Biotechnol.* **2011**, *29*, 205–212. [CrossRef] [PubMed]
13. Liu, K.K.; Cheng, C.L.; Chang, C.C.; Chao, J.I. Biocompatible and detectable carboxylated nanodiamond on human cell. *Nanotechnology* **2007**, *18*. [CrossRef]
14. Berger, M.L. The World of Graphene. In *Nanoengineering: The Skills and Tools Making Technology Invisible*; The Royal Society of Chemistry: London, UK, 2020; pp. 1–60.
15. Malhotra, B.D.; Ali, M.A. (Eds.) Functionalized Carbon Nanomaterials for Biosensors. In *Nanomaterials for Biosensors*; William Andrew Publishing: Norwich, NY, USA, 2018; pp. 75–103.
16. Palmer, B.C.; Phelan-Dickenson, S.J.; DeLouise, L.A. Multi-walled carbon nanotube oxidation dependent keratinocyte cytotoxicity and skin inflammation. *Part. Fibre Toxicol.* **2019**, *16*. [CrossRef]
17. Rout, C.S.; Kumar, A.; Fisher, T.S.; Gautam, U.K.; Bando, Y.; Golberg, D. Synthesis of chemically bonded CNT-graphene heterostructure arrays. *RSC Adv.* **2012**, *2*, 8250–8253. [CrossRef]
18. Zheng, X.T.; Ananthanarayanan, A.; Luo, K.Q.; Chen, P. Glowing Graphene Quantum Dots and Carbon Dots: Properties, Syntheses, and Biological Applications. *Small* **2015**, *11*, 1620–1636. [CrossRef]
19. Terrones, M.; Botello-Mendez, A.R.; Campos-Delgado, J.; Lopez-Urias, F.; Vega-Cantu, Y.I.; Rodriguez-Macias, F.J.; Elias, A.L.; Munoz-Sandoval, E.; Cano-Marquez, A.G.; Charlier, J.C.; et al. Graphene and graphite nanoribbons: Morphology, properties, synthesis, defects and applications. *Nano Today* **2010**, *5*, 351–372. [CrossRef]
20. Karousis, N.; Suarez-Martinez, I.; Ewels, C.P.; Tagmatarchis, N. Structure, Properties, Functionalization, and Applications of Carbon Nanohorns. *Chem. Rev.* **2016**, *116*, 4850–4883. [CrossRef]
21. Heiberg-Andersen, H.; Skjeltorp, A.T.; Sattler, K. Carbon nanocones: A variety of non-crystalline graphite. *J. Non-Cryst. Solids* **2008**, *354*, 5247–5249. [CrossRef]
22. Singh, V.; Joung, D.; Zhai, L.; Das, S.; Khondaker, S.I.; Seal, S. Graphene based materials: Past, present and future. *Prog. Mater. Sci.* **2011**, *56*, 1178–1271. [CrossRef]
23. Yang, K.; Feng, L.Z.; Hong, H.; Cai, W.B.; Liu, Z. Preparation and functionalization of graphene nanocomposites for biomedical applications. *Nat. Protoc.* **2013**, *8*, 2392–2403. [CrossRef] [PubMed]

24. Lacerda, L.; Bianco, A.; Prato, M.; Kostarelos, K. Carbon nanotubes as nanomedicines: From toxicology to pharmacology. *Adv. Drug Deliver. Rev.* **2006**, *58*, 1460–1470. [CrossRef]
25. Bartelmess, J.; Quinn, S.J.; Giordani, S. Carbon nanomaterials: Multi-functional agents for biomedical fluorescence and Raman imaging. *Chem. Soc. Rev.* **2015**, *44*, 4672–4698. [CrossRef] [PubMed]
26. Wong, B.S.; Yoong, S.L.; Jagusiak, A.; Panczyk, T.; Ho, H.K.; Ang, W.H.; Pastorin, G. Carbon nanotubes for delivery of small molecule drugs. *Adv. Drug Deliver. Rev.* **2013**, *65*, 1964–2015. [CrossRef] [PubMed]
27. Goenka, S.; Sant, V.; Sant, S. Graphene-based nanomaterials for drug delivery and tissue engineering. *J. Control. Release* **2014**, *173*, 75–88. [CrossRef] [PubMed]
28. Bianco, A.; Kostarelos, K.; Prato, M. Opportunities and challenges of carbon-based nanomaterials for cancer therapy. *Expert Opin. Drug Del.* **2008**, *5*, 331–342. [CrossRef]
29. Yang, X.Y.; Zhang, X.Y.; Ma, Y.F.; Huang, Y.; Wang, Y.S.; Chen, Y.S. Superparamagnetic graphene oxide-Fe3O4 nanoparticles hybrid for controlled targeted drug carriers. *J. Mater. Chem* **2009**, *19*, 2710–2714. [CrossRef]
30. Cirillo, G.; Curcio, M.; Spizzirri, U.G.; Vittorio, O.; Tucci, P.; Picci, N.; Iemma, F.; Hampel, S.; Nicoletta, F.P. Carbon nanotubes hybrid hydrogels for electrically tunable release of Curcumin. *Eur. Polym. J.* **2017**, *90*, 1–12. [CrossRef]
31. Chen, P.; Wang, Z.Y.; Zong, S.F.; Zhu, D.; Chen, H.; Zhang, Y.Z.; Wu, L.; Cui, Y.P. pH-sensitive nanocarrier based on gold/silver core-shell nanoparticles decorated multi-walled carbon nanotubes for tracing drug release in living cells. *Biosens. Bioelectron.* **2016**, *75*, 446–451. [CrossRef]
32. Cirillo, G.; Curcio, M.; Vittorio, O.; Spizzirri, U.G.; Nicoletta, F.P.; Picci, N.; Hampel, S.; Lemma, F. Dual Stimuli Responsive Gelatin-CNT Hybrid Films as a Versatile Tool for the Delivery of Anionic Drugs. *Macromol. Mater. Eng.* **2016**, *301*, 1537–1547. [CrossRef]
33. Samadian, H.; Salami, M.S.; Jaymand, M.; Azarnezhad, A.; Najafi, M.; Barabadi, H.; Ahmadi, A. Genotoxicity assessment of carbon-based nanomaterials; Have their unique physicochemical properties made them double-edged swords? *Mutat. Res.-Rev. Mutat. Res.* **2020**, *783*. [CrossRef] [PubMed]
34. Poland, C.A.; Duffin, R.; Kinloch, I.; Maynard, A.; Wallace, W.A.H.; Seaton, A.; Stone, V.; Brown, S.; MacNee, W.; Donaldson, K. Carbon nanotubes introduced into the abdominal cavity of mice show asbestos-like pathogenicity in a pilot study. *Nat. Nanotechnol.* **2008**, *3*, 423–428. [CrossRef] [PubMed]
35. Van Berlo, D.; Clift, M.; Albrecht, C.; Schins, R. Carbon nanotubes: An insight into the mechanisms of their potential genotoxicity. *Swiss Med. Wkly.* **2012**, *142*. [CrossRef]
36. Sasidharan, A.; Panchakarla, L.S.; Chandran, P.; Menon, D.; Nair, S.; Rao, C.N.R.; Koyakutty, M. Differential nano-bio interactions and toxicity effects of pristine versus functionalized graphene. *Nanoscale* **2011**, *3*, 2461–2464. [CrossRef] [PubMed]
37. Battigelli, A.; Menard-Moyon, C.; Da Ros, T.; Prato, M.; Bianco, A. Endowing carbon nanotubes with biological and biomedical properties by chemical modifications. *Adv. Drug Deliver. Rev.* **2013**, *65*, 1899–1920. [CrossRef]
38. Shim, M.; Kam, N.W.S.; Chen, R.J.; Li, Y.M.; Dai, H.J. Functionalization of carbon nanotubes for biocompatibility and biomolecular recognition. *Nano Lett.* **2002**, *2*, 285–288. [CrossRef]
39. Karousis, N.; Tagmatarchis, N.; Tasis, D. Current Progress on the Chemical Modification of Carbon Nanotubes. *Chem. Rev.* **2010**, *110*, 5366–5397. [CrossRef]
40. Georgakilas, V.; Tiwari, J.N.; Kemp, K.C.; Perman, J.A.; Bourlinos, A.B.; Kim, K.S.; Zboril, R. Noncovalent Functionalization of Graphene and Graphene Oxide for Energy Materials, Biosensing, Catalytic, and Biomedical Applications. *Chem. Rev.* **2016**, *116*, 5464–5519. [CrossRef]
41. Antonucci, A.; Kupis-Rozmyslowicz, J.; Boghossian, A.A. Noncovalent Protein and Peptide Functionalization of Single-Walled Carbon Nanotubes for Biodelivery and Optical Sensing Applications. *ACS Appl. Mater. Inter.* **2017**, *9*, 11321–11331. [CrossRef]
42. Nava, A.C.; Cojoc, M.; Peitzsch, C.; Cirillo, G.; Kurth, I.; Fuessel, S.; Erdmann, K.; Kunhardt, D.; Vittorio, O.; Hampel, S.; et al. Development of novel radiochemotherapy approaches targeting prostate tumor progenitor cells using nanohybrids. *Int. J. Cancer* **2015**, *137*, 2492–2503. [CrossRef]
43. Wu, H.C.; Chang, X.L.; Liu, L.; Zhao, F.; Zhao, Y.L. Chemistry of carbon nanotubes in biomedical applications. *J. Mater. Chem.* **2010**, *20*, 1036–1052. [CrossRef]
44. Wang, X.L.; Shi, G.Q. An introduction to the chemistry of graphene. *Phys. Chem. Chem. Phys.* **2015**, *17*, 28484–28504. [CrossRef]

45. Matson, M.L.; Villa, C.H.; Ananta, J.S.; Law, J.J.; Scheinberg, D.A.; Wilson, L.J. Encapsulation of alpha-Particle-Emitting Ac-225(3+) Ions Within Carbon Nanotubes. *J. Nucl. Med.* **2015**, *56*, 897–900. [CrossRef] [PubMed]
46. Chronopoulos, D.D.; Bakandritsos, A.; Pykal, M.; Zboril, R.; Otyepka, M. Chemistry, properties, and applications of fluorographene. *Appl. Mater. Today* **2017**, *9*, 60–70. [CrossRef] [PubMed]
47. Taylor, R. Fluorinated fullerenes. *Chem.-Eur. J.* **2001**, *7*, 4074–4083. [CrossRef]
48. Mohammadi, S.; Kolahdouz, Z.; Mohajerzadeh, S. Hydrogenation-assisted unzipping of carbon nanotubes to realize graphene nano-sheets. *J. Mater. Chem. C* **2013**, *1*, 1309–1316. [CrossRef]
49. Wang, Y.; Wang, C. Self-assembly of graphene sheets actuated by surface topological defects: Toward the fabrication of novel nanostructures and drug delivery devices. *Appl. Surf. Sci.* **2020**, *505*. [CrossRef]
50. Schur, D.V.; Zaginaichenko, S.Y.; Veziroğlu, T.N.; Javadov, N.F. The Peculiarities of Hydrogenation of Fullerene Molecules C60 and Their Transformation. In *Black Sea Energy Resource Development and Hydrogen Energy Problems*; Veziroğlu, A., Tsitskishvili, M., Eds.; Springer: Dordrecht, The Netherlands, 2013; pp. 191–204.
51. Cheng, Q.S.; Blais, M.O.; Harris, G.; Jabbarzadeh, E. PLGA-Carbon Nanotube Conjugates for Intercellular Delivery of Caspase-3 into Osteosarcoma Cells. *PLoS ONE* **2013**, *8*. [CrossRef]
52. Chen, S.; Liu, J.W.; Chen, M.L.; Chen, X.W.; Wang, J.H. Unusual emission transformation of graphene quantum dots induced by self-assembled aggregation. *Chem. Commun.* **2012**, *48*, 7637–7639. [CrossRef]
53. Chua, C.K.; Sofer, Z.; Simek, P.; Jankovsky, O.; Klimova, K.; Bakardjieva, S.; Kuckova, S.H.; Pumera, M. Synthesis of Strongly Fluorescent Graphene Quantum Dots by Cage-Opening Buckminsterfullerene. *ACS Nano* **2015**, *9*, 2548–2555. [CrossRef]
54. Pippa, N.; Chronopoulos, D.D.; Stellas, D.; Fernandez-Pacheco, R.; Arenal, R.; Demetzos, C.; Tagmatarchis, N. Design and development of multi-walled carbon nanotube-liposome drug delivery platforms. *Int. J. Pharmaceut.* **2017**, *528*, 429–439. [CrossRef] [PubMed]
55. Dreyer, D.R.; Park, S.; Bielawski, C.W.; Ruoff, R.S. The chemistry of graphene oxide. *Chem. Soc. Rev.* **2010**, *39*, 228–240. [CrossRef] [PubMed]
56. Afreen, S.; Kokubo, K.; Muthoosamy, K.; Manickam, S. Hydration or hydroxylation: Direct synthesis of fullerenol from pristine fullerene [C-60] via acoustic cavitation in the presence of hydrogen peroxide. *RSC Adv.* **2017**, *7*, 31930–31939. [CrossRef]
57. Beckler, B.; Cowan, A.; Farrar, N.; Murawski, A.; Robinson, A.; Diamanduros, A.; Scarpinato, K.; Sittaramane, V.; Quirino, R.L. Microwave Heating of Antibody-functionalized Carbon Nanotubes as a Feasible Cancer Treatment. *Biomed. Phys. Eng. Expr.* **2018**, *4*. [CrossRef]
58. Campbell, E.; Hasan, M.T.; Pho, C.; Callaghan, K.; Akkaraju, G.R.; Naumov, A.V. Graphene Oxide as a Multifunctional Platform for Intracellular Delivery, Imaging, and Cancer Sensing. *Sci. Rep.* **2019**, *9*. [CrossRef] [PubMed]
59. Sabirov, D.S.; Khursan, S.L.; Bulgakov, R.G. 1,3-Dipolar addition reactions to fullerenes: The role of the local curvature of carbon surface. *Russ. Chem. B+* **2008**, *57*, 2520–2525. [CrossRef]
60. Viswanathan, G.; Chakrapani, N.; Yang, H.C.; Wei, B.Q.; Chung, H.S.; Cho, K.W.; Ryu, C.Y.; Ajayan, P.M. Single-step in situ synthesis of polymer-grafted single-wall nanotube composites. *J. Am. Chem. Soc.* **2003**, *125*, 9258–9259. [CrossRef]
61. Georgakilas, V.; Otyepka, M.; Bourlinos, A.B.; Chandra, V.; Kim, N.; Kemp, K.C.; Hobza, P.; Zboril, R.; Kim, K.S. Functionalization of Graphene: Covalent and Non-Covalent Approaches, Derivatives and Applications. *Chem. Rev.* **2012**, *112*, 6156–6214. [CrossRef] [PubMed]
62. Lin, H.S.; Matsuo, Y. Functionalization of [60]fullerene through fullerene cation intermediates. *Chem. Commun.* **2018**, *54*, 11244–11259. [CrossRef] [PubMed]
63. Kooi, S.E.; Schlecht, U.; Burghard, M.; Kern, K. Electrochemical modification of single carbon nanotubes. *Angew. Chem. Int. Edit.* **2002**, *41*, 1353–1355. [CrossRef]
64. Qi, M.; Zhang, Y.; Cao, C.M.; Zhang, M.X.; Liu, S.H.; Liu, G.Z. Decoration of Reduced Graphene Oxide Nanosheets with Aryldiazonium Salts and Gold Nanoparticles toward a Label-Free Amperometric Immunosensor for Detecting Cytokine Tumor Necrosis Factor-alpha in Live Cells. *Anal. Chem.* **2016**, *88*, 9614–9621. [CrossRef]
65. Flavin, K.; Chaur, M.N.; Echegoyen, L.; Giordani, S. Functionalization of Multilayer Fullerenes (Carbon Nano-Onions) using Diazonium Compounds and "Click" Chemistry. *Org. Lett.* **2010**, *12*, 840–843. [CrossRef] [PubMed]

66. Koromilas, N.D.; Lainioti, G.C.; Gialeli, C.; Barbouri, D.; Kouravelou, K.B.; Karamanos, N.K.; Voyiatzis, G.A.; Kallitsis, J.K. Preparation and Toxicological Assessment of Functionalized Carbon Nanotube-Polymer Hybrids. *PLoS One* **2014**, *9*. [CrossRef] [PubMed]
67. Vusa, C.S.R.; Venkatesan, M.; Aneesh, K.; Berchmans, S.; Arumugam, P. Tactical tuning of the surface and interfacial properties of graphene: A Versatile and rational electrochemical approach. *Sci. Rep.* **2017**, *7*. [CrossRef] [PubMed]
68. Ramirez-Calera, I.J.; Meza-Laguna, V.; Gromovoy, T.Y.; Chavez-Uribe, M.I.; Basiuk, V.A.; Basiuk, E.V. Solvent-free functionalization of fullerene C-60 and pristine multi-walled carbon nanotubes with aromatic amines. *Appl. Surf. Sci.* **2015**, *328*, 45–62. [CrossRef]
69. Homenick, C.M.; Lawson, G.; Adronov, A. Polymer grafting of carbon nanotubes using living free-radical polymerization. *Polym. Rev.* **2007**, *47*, 265–290. [CrossRef]
70. Servant, A.; Bianco, A.; Prato, M.; Kostarelos, K. Graphene for multi-functional synthetic biology: The last 'zeitgeist' in nanomedicine. *Bioorg. Med. Chem. Lett.* **2014**, *24*, 1638–1649. [CrossRef]
71. Hasanzadeh, A.; Khataee, A.; Zarei, M.; Zhang, Y.F. Two-electron oxygen reduction on fullerene C-60-carbon nanotubes covalent hybrid as a metal-free electrocatalyst. *Sci. Rep.* **2019**, *9*. [CrossRef]
72. Ismaili, H.; Lagugne-Labarthet, F.; Workentin, M.S. Covalently Assembled Gold Nanoparticle-Carbon Nanotube Hybrids via a Photoinitiated Carbene Addition Reaction. *Chem. Mater.* **2011**, *23*, 1519–1525. [CrossRef]
73. Ismaili, H.; Geng, D.S.; Sun, A.X.L.; Kantzas, T.T.; Workentin, M.S. Light-Activated Covalent Formation of Gold Nanoparticle Graphene and Gold Nanoparticle-Glass Composites. *Langmuir* **2011**, *27*, 13261–13268. [CrossRef]
74. Lorbach, A.; Maverick, E.; Carreras, A.; Alemany, P.; Wu, G.; Garcia-Garibay, M.A.; Bazan, G.C. A fullerene-carbene adduct as a crystalline molecular rotor: Remarkable behavior of a spherically-shaped rotator. *Phys. Chem. Chem. Phys.* **2014**, *16*, 12980–12986. [CrossRef] [PubMed]
75. Boncel, S.; Pluta, A.; Skonieczna, M.; Gondela, A.; Maciejewska, B.; Herman, A.P.; Jedrysiak, R.G.; Budniok, S.; Komedera, K.; Blachowski, A.; et al. Hybrids of Iron-Filled Multiwall Carbon Nanotubes and Anticancer Agents as Potential Magnetic Drug Delivery Systems: In Vitro Studies against Human Melanoma, Colon Carcinoma, and Colon Adenocarcinoma. *J. Nanomater.* **2017**. [CrossRef]
76. Quintana, M.; Spyrou, K.; Grzelczak, M.; Browne, W.R.; Rudolf, P.; Prato, M. Functionalization of Graphene via 1,3-Dipolar Cycloaddition. *ACS Nano* **2010**, *4*, 3527–3533. [CrossRef] [PubMed]
77. Nakahodo, T.; Okada, M.; Morita, H.; Yoshimura, T.; Ishitsuka, M.O.; Tsuchiya, T.; Maeda, Y.; Fujihara, H.; Akasaka, T.; Gao, X.; et al. [2+1] cycloaddition of nitrene onto C(60) revisited: Interconversion between an aziridinofullerene and an azafulleroid. *Angew. Chem. Int. Edit.* **2008**, *47*, 1298–1300. [CrossRef] [PubMed]
78. Samori, C.; Ali-Boucetta, H.; Sainz, R.; Guo, C.; Toma, F.M.; Fabbro, C.; da Ros, T.; Prato, M.; Kostarelos, K.; Bianco, A. Enhanced anticancer activity of multi-walled carbon nanotube-methotrexate conjugates using cleavable linkers. *Chem. Commun.* **2010**, *46*, 1494–1496. [CrossRef] [PubMed]
79. Bekiari, V.; Karakassides, A.; Georgitsopoulou, S.; Kouloumpis, A.; Gournis, D.; Georgakilas, V. Self-assembly of one-side-functionalized graphene nanosheets in bilayered superstructures for drug delivery. *J. Mater. Sci.* **2018**, *53*, 11167–11175. [CrossRef]
80. Pacor, S.; Grillo, A.; Dordevic, L.; Zorzet, S.; Lucafo, M.; Da Ros, T.; Prato, M.; Sava, G. Effects of Two Fullerene Derivatives on Monocytes and Macrophages. *Biomed. Res. Int.* **2015**. [CrossRef]
81. Irannejad, S.; Amini, M.; Modanlookordi, M.; Shokrzadeh, M.; Irannejad, H. Preparation of Diaminedicarboxyplatinum (II) Functionalized Single-Wall Carbon Nanotube via Bingel Reaction as a Novel Cytotoxic Agent. *Iran. J. Pharm. Res.* **2016**, *15*, 753–762.
82. Stergiou, A.; Pagona, G.; Tagmatarchis, N. Donor-acceptor graphene-based hybrid materials facilitating photo-induced electron-transfer reactions. *Beilstein J. Nanotech.* **2014**, *5*, 1580–1589. [CrossRef]
83. Biglova, Y.N.; Mustafin, A.G. Nucleophilic cyclopropanation of [60]fullerene by the addition-elimination mechanism. *RSC Adv.* **2019**, *9*, 22428–22498. [CrossRef]
84. Ondera, T.J.; Hamme, A.T. A gold nanopopcorn attached single-walled carbon nanotube hybrid for rapid detection and killing of bacteria. *J. Mater. Chem. B* **2014**, *2*, 7534–7543. [CrossRef] [PubMed]
85. Barrejon, M.; Gomez-Escalonilla, M.J.; Fierro, J.L.G.; Prieto, P.; Carrillo, J.R.; Rodriguez, A.M.; Abellan, G.; Lopez-Escalante, M.C.; Gabas, M.; Lopez-Navarrete, J.T.; et al. Modulation of the exfoliated graphene work function through cycloaddition of nitrile imines. *Phys. Chem. Chem. Phys.* **2016**, *18*, 29582–29590. [CrossRef]

86. Sugawara, Y.; Jasinski, N.; Kaupp, M.; Welle, A.; Zydziak, N.; Blasco, E.; Barner-Kowollik, C. Light-driven nitrile imine-mediated tetrazole-ene cycloaddition as a versatile platform for fullerene conjugation. *Chem. Commun.* **2015**, *51*, 13000–13003. [CrossRef] [PubMed]
87. Cao, X.T.; Patil, M.P.; Phan, Q.T.; Le, C.M.Q.; Ahn, B.-H.; Kim, G.-D.; Lim, K.T. Green and direct functionalization of poly (ethylene glycol) grafted polymers onto single walled carbon nanotubes: Effective nanocarrier for doxorubicin delivery. *J. Ind. Eng. Chem.* **2020**, *83*, 173–180. [CrossRef]
88. Yuan, J.C.; Chen, G.H.; Weng, W.G.; Xu, Y.Z. One-step functionalization of graphene with cyclopentadienyl-capped macromolecules via Diels-Alder "click" chemistry. *J. Mater. Chem.* **2012**, *22*, 7929–7936. [CrossRef]
89. Tsuda, M.; Ishida, T.; Nogami, T.; Kurono, S.; Ohashi, M. Isolation and Characterization of Diels-Alder Adducts of C-60 with Anthracene and Cyclopentadiene. *J. Chem. Soc. Chem. Comm.* **1993**, 1296–1298. [CrossRef]
90. Pastorin, G. Crucial Functionalizations of Carbon Nanotubes for Improved Drug Delivery: A Valuable Option? *Pharm. Res.-Dordr.* **2009**, *26*, 746–769. [CrossRef] [PubMed]
91. Ziegler, K.J.; Gu, Z.N.; Peng, H.Q.; Flor, E.L.; Hauge, R.H.; Smalley, R.E. Controlled oxidative cutting of single-walled carbon nanotubes. *J. Am. Chem. Soc.* **2005**, *127*, 1541–1547. [CrossRef]
92. Savage, T.; Bhattacharya, S.; Sadanadan, B.; Gaillard, J.; Tritt, T.M.; Sun, Y.P.; Wu, Y.; Nayak, S.; Car, R.; Marzari, N.; et al. Photoinduced oxidation of carbon nanotubes. *J. Phys.: Condens. Matter* **2003**, *15*, 5915–5921. [CrossRef]
93. Felten, A.; Bittencourt, C.; Pireaux, J.J. Gold clusters on oxygen plasma functionalized carbon nanotubes: XPS and TEM studies. *Nanotechnology* **2006**, *17*, 1954–1959. [CrossRef]
94. Datsyuk, V.; Kalyva, M.; Papagelis, K.; Parthenios, J.; Tasis, D.; Siokou, A.; Kallitsis, I.; Galiotis, C. Chemical oxidation of multiwalled carbon nanotubes. *Carbon* **2008**, *46*, 833–840. [CrossRef]
95. Jiang, L.Q.; Gao, L.; Sun, J. Production of aqueous colloidal dispersions of carbon nanotubes. *J. Colloid Interface Sci.* **2003**, *260*, 89–94. [CrossRef]
96. Rosca, I.D.; Watari, F.; Uo, M.; Akaska, T. Oxidation of multiwalled carbon nanotubes by nitric acid. *Carbon* **2005**, *43*, 3124–3131. [CrossRef]
97. Witzmann, F.A.; Monteiro-Riviere, N.A. Multi-walled carbon nanotube exposure alters protein expression in human keratinocytes. *Nanomedicine* **2006**, *2*, 158–168. [CrossRef] [PubMed]
98. Bussy, C.; Hadad, C.; Prato, M.; Bianco, A.; Kostarelos, K. Intracellular degradation of chemically functionalized carbon nanotubes using a long-term primary microglial culture model. *Nanoscale* **2016**, *8*, 590–601. [CrossRef] [PubMed]
99. Tagmatarchis, N.; Prato, M. Functionalization of carbon nanotubes via 1,3-dipolar cycloadditions. *J. Mater. Chem.* **2004**, *14*, 437–439. [CrossRef]
100. Okotrub, A.V.; Maksimova, N.; Duda, T.A.; Kudashov, A.G.; Shubin, Y.V.; Su, D.S.; Pazhetnov, E.M.; Boronin, A.I.; Bulusheva, L.G. Fluorination of CNx nanotubes. *Fuller. Nanotub. Carbon Nanostructures* **2004**, *12*, 99–104. [CrossRef]
101. Struzzi, C.; Scardamaglia, M.; Hemberg, A.; Petaccia, L.; Colomer, J.F.; Snyders, R.; Bittencourt, C. Plasma fluorination of vertically aligned carbon nanotubes: Functionalization and thermal stability. *Beilstein J. Nanotech.* **2015**, *6*, 2263–2271. [CrossRef]
102. Li, Y.; Chen, Y.F.; Feng, Y.Y.; Zhao, S.L.; Lu, P.; Yuan, X.Y.; Feng, W. Progress of synthesizing methods and properties of fluorinated carbon nanotubes. *Sci. China Technol. Sci.* **2010**, *53*, 1225–1233. [CrossRef]
103. Kuila, T.; Bose, S.; Mishra, A.K.; Khanra, P.; Kim, N.H.; Lee, J.H. Chemical functionalization of graphene and its applications. *Prog. Mater. Sci.* **2012**, *57*, 1061–1105. [CrossRef]
104. Makharza, S.; Cirillo, G.; Bachmatiuk, A.; Vittorio, O.; Mendes, R.G.; Oswald, S.; Hampel, S.; Rummeli, M.H. Size-dependent nanographene oxide as a platform for efficient carboplatin release. *J. Mater. Chem. B* **2013**, *1*, 6107–6114. [CrossRef] [PubMed]
105. Marcano, D.C.; Kosynkin, D.V.; Berlin, J.M.; Sinitskii, A.; Sun, Z.Z.; Slesarev, A.; Alemany, L.B.; Lu, W.; Tour, J.M. Improved Synthesis of Graphene Oxide. *ACS Nano* **2010**, *4*, 4806–4814. [CrossRef] [PubMed]
106. Zhang, L.M.; Xia, J.G.; Zhao, Q.H.; Liu, L.W.; Zhang, Z.J. Functional Graphene Oxide as a Nanocarrier for Controlled Loading and Targeted Delivery of Mixed Anticancer Drugs. *Small* **2010**, *6*, 537–544. [CrossRef] [PubMed]
107. Sahu, A.; Choi, W.I.; Lee, J.H.; Tae, G. Graphene oxide mediated delivery of methylene blue for combined photodynamic and photothermal therapy. *Biomaterials* **2013**, *34*, 6239–6248. [CrossRef] [PubMed]

108. Xu, Z.Y.; Wang, S.; Li, Y.J.; Wang, M.W.; Shi, P.; Huang, X.Y. Covalent Functionalization of Graphene Oxide with Biocompatible Poly(ethylene glycol) for Delivery of Paclitaxel. *ACS Appl. Mater. Inter.* **2014**, *6*, 17268–17276. [CrossRef]
109. Stankovich, S.; Dikin, D.A.; Piner, R.D.; Kohlhaas, K.A.; Kleinhammes, A.; Jia, Y.; Wu, Y.; Nguyen, S.T.; Ruoff, R.S. Synthesis of graphene-based nanosheets via chemical reduction of exfoliated graphite oxide. *Carbon* **2007**, *45*, 1558–1565. [CrossRef]
110. Cherian, R.S.; Anju, S.; Paul, W.; Sabareeswaran, A.; Mohanan, P.V. Organ distribution and biological compatibility of surface-functionalized reduced graphene oxide. *Nanotechnology* **2020**, *31*. [CrossRef]
111. Darabdhara, G.; Das, M.R.; Turcheniuk, V.; Turcheniuk, K.; Zaitsev, V.; Boukherroub, R.; Szunerits, S. Reduced graphene oxide nanosheets decorated with AuPd bimetallic nanoparticles: A multifunctional material for photothermal therapy of cancer cells. *J. Mater. Chem. B* **2015**, *3*, 8366–8374. [CrossRef]
112. Zainuddin, M.F.; Raikhan, N.H.N.; Othman, N.H.; Abdullah, W.F.H. Synthesis of reduced Graphene Oxide (rGO) using different treatments of Graphene Oxide (GO). *IOP Conf. Ser.-Mat. Sci.* **2018**. [CrossRef]
113. Kalluri, A.; Debnath, D.; Dharmadhikari, B.; Patra, P. Graphene Quantum Dots: Synthesis and Applications. In *Methods in Enzymology*; Kumar, C.V., Ed.; Academic Press: Cambridge, MA, USA, 2018; Volume 609, pp. 335–354.
114. Zhao, M.L. Direct Synthesis of Graphene Quantum Dots with Different Fluorescence Properties by Oxidation of Graphene Oxide Using Nitric Acid. *Appl. Sci.* **2018**, *8*. [CrossRef]
115. Pan, D.Y.; Guo, L.; Zhang, J.C.; Xi, C.; Xue, Q.; Huang, H.; Li, J.H.; Zhang, Z.W.; Yu, W.J.; Chen, Z.W.; et al. Cutting sp(2) clusters in graphene sheets into colloidal graphene quantum dots with strong green fluorescence. *J. Mater. Chem.* **2012**, *22*, 3314–3318. [CrossRef]
116. Milane, L.; Ganesh, S.; Shah, S.; Duan, Z.F.; Amiji, M. Multi-modal strategies for overcoming tumor drug resistance: Hypoxia, the Warburg effect, stem cells, and multifunctional nanotechnology. *J. Control. Release* **2011**, *155*, 237–247. [CrossRef] [PubMed]
117. Yalcin, S.; Ozluer, O.; Gunduz, U. Nanoparticle-based drug delivery in cancer: The role of cell membrane structures. *Ther. Deliv.* **2016**, *7*, 773–781. [CrossRef] [PubMed]
118. Dong, X.W.; Mumper, R.J. Nanomedicinal strategies to treat multidrug-resistant tumors: Current progress. *Nanomedicine* **2010**, *5*, 597–615. [CrossRef] [PubMed]
119. Katiyar, S.S.; Muntimadugu, E.; Rafeeqi, T.A.; Domb, A.J.; Khan, W. Co-delivery of rapamycin- and piperine-loaded polymeric nanoparticles for breast cancer treatment. *Drug Deliv.* **2016**, *23*, 2608–2616. [CrossRef]
120. Alibert-Franco, S.; Pradines, B.; Mahamoud, A.; Davin-Regli, A.; Pages, J.M. Efflux Mechanism, an Attractive Target to Combat Multidrug Resistant Plasmodium falciparum and Pseudomonas aeruginosa. *Curr. Med. Chem.* **2009**, *16*, 301–317. [CrossRef]
121. Gillet, J.P.; Gottesman, M.M. Advances in the Molecular Detection of ABC Transporters Involved in Multidrug Resistance in Cancer. *Curr. Pharm. Biotechnol.* **2011**, *12*, 686–692. [CrossRef]
122. Yu, M.; Ocana, A.; Tannock, I.F. Reversal of ATP-binding cassette drug transporter activity to modulate chemoresistance: Why has it failed to provide clinical benefit? *Cancer Metast. Rev.* **2013**, *32*, 211–227. [CrossRef]
123. Fromm, M.F. Importance of P-glycoprotein at blood-tissue barriers. *Trends Pharmacol. Sci.* **2004**, *25*, 423–429. [CrossRef]
124. Wang, Z.J.; Xu, Y.H.; Meng, X.N.; Watari, F.M.; Liu, H.D.; Chen, X. Suppression of c-Myc is involved in multi-walled carbon nanotubes' down-regulation of ATP-binding cassette transporters in human colon adenocarcinoma cells. *Toxicol. Appl. Pharm.* **2015**, *282*, 42–51. [CrossRef]
125. Fabbro, C.; Ali-Boucetta, H.; Da Ros, T.; Kostarelos, K.; Bianco, A.; Prato, M. Targeting carbon nanotubes against cancer. *Chem. Commun.* **2012**, *48*, 3911–3926. [CrossRef] [PubMed]
126. Shityakov, S.; Förster, C. Multidrug resistance protein P-gp interaction with nanoparticles (fullerenes and carbon nanotube) to assess their drug delivery potential: A theoretical molecular docking study. *Int. J. Comput. Biol. Drug Des.* **2013**, *6*, 343–357. [CrossRef] [PubMed]
127. Cheng, J.P.; Meziani, M.J.; Sun, Y.P.; Cheng, S.H. Poly(ethylene glycol)-conjugated multi-walled carbon nanotubes as an efficient drug carrier for overcoming multidrug resistance. *Toxicol. Appl. Pharm.* **2011**, *250*, 184–193. [CrossRef]
128. Kumar, M.; Sharma, G.; Misra, C.; Kumar, R.; Singh, B.; Katare, O.P.; Raza, K. N-desmethyl tamoxifen and quercetin-loaded multiwalled CNTs: A synergistic approach to overcome MDR in cancer cells. *Mat. Sci. Eng. C* **2018**, *89*, 274–282. [CrossRef]

129. Luo, C.; Li, Y.F.; Guo, L.J.; Zhang, F.W.; Liu, H.; Zhang, J.L.; Zheng, J.; Zhang, J.Y.; Guo, S.W. Graphene Quantum Dots Downregulate Multiple Multidrug-Resistant Genes via Interacting with Their C-Rich Promoters. *Adv. Healthc. Mater.* **2017**, *6*. [CrossRef] [PubMed]
130. Miao, Y.Y.; Zhang, H.X.; Pan, Y.B.; Ren, J.; Ye, M.M.; Xia, F.F.; Huang, R.; Lin, Z.H.; Jiang, S.; Zhang, Y.; et al. Single-walled carbon nanotube: One specific inhibitor of cancer stem cells in osteosarcoma upon downregulation of the TGF beta 1 signaling. *Biomaterials* **2017**, *149*, 29–40. [CrossRef]
131. Da Silva, C.G.; Peters, G.J.; Ossendorp, F.; Cruz, L.J. The potential of multi-compound nanoparticles to bypass drug resistance in cancer. *Cancer Chemother. Pharmacol.* **2017**, *80*, 881–894. [CrossRef]
132. Salvioni, L.; Rizzuto, M.A.; Bertolini, J.A.; Pandolfi, L.; Colombo, M.; Prosperi, D. Thirty Years of Cancer Nanomedicine: Success, Frustration, and Hope. *Cancers* **2019**, *11*. [CrossRef]
133. Kumawat, M.K.; Thakur, M.; Gurung, R.B.; Srivastava, R. Graphene Quantum Dots for Cell Proliferation, Nucleus Imaging, and Photoluminescent Sensing Applications. *Sci. Rep.* **2017**, *7*. [CrossRef]
134. Mu, Q.X.; Broughton, D.L.; Yan, B. Endosomal Leakage and Nuclear Translocation of Multiwalled Carbon Nanotubes: Developing a Model for Cell Uptake. *Nano Lett.* **2009**, *9*, 4370–4375. [CrossRef]
135. Shi, X.H.; von dem Bussche, A.; Hurt, R.H.; Kane, A.B.; Gao, H.J. Cell entry of one-dimensional nanomaterials occurs by tip recognition and rotation. *Nat. Nanotechnol.* **2011**, *6*, 714–719. [CrossRef] [PubMed]
136. Kam, N.W.S.; Jessop, T.C.; Wender, P.A.; Dai, H.J. Nanotube molecular transporters: Internalization of carbon nanotube-protein conjugates into mammalian cells. *J. Am. Chem Soc.* **2004**, *126*, 6850–6851. [CrossRef] [PubMed]
137. Costa, P.M.; Bourgognon, M.; Wang, J.T.W.; Al-Jamal, K.T. Functionalised carbon nanotubes: From intracellular uptake and cell-related toxicity to systemic brain delivery. *J. Control. Release* **2016**, *241*, 200–219. [CrossRef] [PubMed]
138. Huang, J.; Zong, C.; Shen, H.; Liu, M.; Chen, B.A.; Ren, B.; Zhang, Z.J. Mechanism of Cellular Uptake of Graphene Oxide Studied by Surface-Enhanced Raman Spectroscopy. *Small* **2012**, *8*, 2577–2584. [CrossRef] [PubMed]
139. Hashimoto, A.; Yamanaka, T.; Takamura-Enya, T. Synthesis of novel fluorescently labeled water-soluble fullerenes and their application to its cellar uptake and distribution properties. *J. Nanopart. Res.* **2017**, *19*. [CrossRef]
140. Mahajan, S.; Patharkar, A.; Kuche, K.; Maheshwari, R.; Deb, P.K.; Kalia, K.; Tekade, R.K. Functionalized carbon nanotubes as emerging delivery system for the treatment of cancer. *Int. J. Pharmaceut.* **2018**, *548*, 540–558. [CrossRef]
141. Gong, P.W.; Zhang, L.; Yuan, X.A.; Liu, X.C.; Diao, X.L.; Zhao, Q.; Tian, Z.Z.; Sun, J.; Liu, Z.; You, J.M. Multifunctional fluorescent PEGylated fluorinated graphene for targeted drug delivery: An experiment and DFT study. *Dyes Pigments* **2019**, *162*, 573–582. [CrossRef]
142. Mahmood, M.; Xu, Y.; Dantuluri, V.; Mustafa, T.; Zhang, Y.; Karmakar, A.; Casciano, D.; Ali, S.; Biris, A. Carbon nanotubes enhance the internalization of drugs by cancer cells and decrease their chemoresistance to cytostatics. *Nanotechnology* **2013**, *24*. [CrossRef]
143. Wu, P.P.; Li, S.; Zhang, H.J. Design real-time reversal of tumor multidrug resistance cleverly with shortened carbon nanotubes. *Drug Des. Dev. Ther.* **2014**, *8*, 2431–2438. [CrossRef]
144. Alizadeh, D.; White, E.E.; Sanchez, T.C.; Liu, S.N.; Zhang, L.Y.; Badie, B.; Berlin, J.M. Immunostimulatory CpG on Carbon Nanotubes Selectively Inhibits Migration of Brain Tumor Cells. *Bioconjugate Chem.* **2018**, *29*, 1659–1668. [CrossRef]
145. Lin, K.C.; Lin, M.W.; Hsu, M.N.; Guan, Y.C.; Chao, Y.C.; Tuan, H.Y.; Chiang, C.S.; Hu, Y.C. Graphene oxide sensitizes cancer cells to chemotherapeutics by inducing early autophagy events, promoting nuclear trafficking and necrosis. *Theranostics* **2018**, *8*, 2477–2487. [CrossRef] [PubMed]
146. Li, Y.L.; Gao, X.N.; Yu, Z.Z.; Liu, B.; Pan, W.; Li, N.; Tang, B. Reversing Multidrug Resistance by Multiplexed Gene Silencing for Enhanced Breast Cancer Chemotherapy. *ACS Appl. Mater. Inter.* **2018**, *10*, 15461–15466. [CrossRef] [PubMed]
147. Guven, A.; Rusakova, I.A.; Lewis, M.T.; Wilson, L.J. Cisplatin@US-tube carbon nanocapsules for enhanced chemotherapeutic delivery. *Biomaterials* **2012**, *33*, 1455–1461. [CrossRef] [PubMed]
148. Guven, A.; Villares, G.J.; Hilsenbeck, S.G.; Lewis, A.; Landua, J.D.; Dobrolecki, L.E.; Wilson, L.J.; Lewis, M.T. Carbon nanotube capsules enhance the in vivo efficacy of cisplatin. *Acta Biomater.* **2017**, *58*, 466–478. [CrossRef] [PubMed]

149. Muzi, L.; Menard-Moyon, C.; Russier, J.; Li, J.; Chin, C.F.; Ang, W.H.; Pastorin, G.; Risuleo, G.; Bianco, A. Diameter-dependent release of a cisplatin pro-drug from small and large functionalized carbon nanotubes. *Nanoscale* **2015**, *7*, 5383–5394. [CrossRef]
150. Sui, X.; Luo, C.; Wang, C.; Zhang, F.W.; Zhang, J.Y.; Guo, S.W. Graphene quantum dots enhance anticancer activity of cisplatin via increasing its cellular and nuclear uptake. *Nanomedicine* **2016**, *12*, 1997–2006. [CrossRef]
151. Wei, Z.; Yin, X.T.; Cai, Y.; Xu, W.G.; Song, C.H.; Wang, Y.F.; Zhang, J.W.; Kang, A.; Wang, Z.Y.; Han, W. Antitumor effect of a Pt-loaded nanocomposite based on graphene quantum dots combats hypoxia-induced chemoresistance of oral squamous cell carcinoma. *Int. J. Nanomed.* **2018**, *13*, 1505–1524. [CrossRef]
152. Wang, C.; Wu, C.Y.; Zhou, X.J.; Han, T.; Xin, X.Z.; Wu, J.Y.; Zhang, J.Y.; Guo, S.W. Enhancing Cell Nucleus Accumulation and DNA Cleavage Activity of Anti-Cancer Drug via Graphene Quantum Dots. *Sci. Rep.* **2013**, *3*. [CrossRef]
153. Wu, J.; Wang, Y.S.; Yang, X.Y.; Liu, Y.Y.; Yang, J.R.; Yang, R.; Zhang, N. Graphene oxide used as a carrier for adriamycin can reverse drug resistance in breast cancer cells. *Nanotechnology* **2012**, *23*. [CrossRef]
154. Jin, R.; Ji, X.J.; Yang, Y.X.; Wang, H.F.; Cao, A.N. Self-Assembled Graphene-Dextran Nanohybrid for Killing Drug-Resistant Cancer Cells. *ACS Appl. Mater. Inter.* **2013**, *5*, 7181–7189. [CrossRef]
155. Zhang, Q.; Chi, H.R.; Tang, M.Z.; Chen, J.B.; Li, G.L.; Liu, Y.S.; Liu, B. Mixed surfactant modified graphene oxide nanocarriers for DOX delivery to cisplatin-resistant human ovarian carcinoma cells. *RSC Adv.* **2016**, *6*, 87258–87269. [CrossRef]
156. Wu, C.H.; Cao, C.; Kim, J.H.; Hsu, C.H.; Wanebo, H.J.; Bowen, W.D.; Xu, J.; Marshall, J. Trojan-Horse Nanotube On-Command Intracellular Drug Delivery. *Nano Lett.* **2012**, *12*, 5475–5480. [CrossRef] [PubMed]
157. Yao, H.J.; Sun, L.; Liu, Y.; Jiang, S.; Pu, Y.Z.; Li, J.C.; Zhang, Y.G. Monodistearoylphosphatidylethanolamine-hyaluronic acid functionalization of single-walled carbon nanotubes for targeting intracellular drug delivery to overcome multidrug resistance of cancer cells. *Carbon* **2016**, *96*, 362–376. [CrossRef]
158. Zhi, F.; Dong, H.F.; Jia, X.F.; Guo, W.J.; Lu, H.T.; Yang, Y.L.; Ju, H.X.; Zhang, X.J.; Hu, Y.Q. Functionalized Graphene Oxide Mediated Adriamycin Delivery and miR-21 Gene Silencing to Overcome Tumor Multidrug Resistance In Vitro. *PLoS One* **2013**, *8*. [CrossRef]
159. Lu, C.H.; Zhu, C.L.; Li, J.; Liu, J.J.; Chen, X.; Yang, H.H. Using graphene to protect DNA from cleavage during cellular delivery. *Chem. Commun.* **2010**, *46*, 3116–3118. [CrossRef]
160. Kolosnjaj-Tabi, J.; Hartman, K.B.; Boudjemaa, S.; Ananta, J.S.; Morgant, G.; Szwarc, H.; Wilson, L.J.; Moussa, F. In Vivo Behavior of Large Doses of Ultrashort and Full-Length Single-Walled Carbon Nanotubes after Oral and Intraperitoneal Administration to Swiss Mice. *ACS Nano* **2010**, *4*, 1481–1492. [CrossRef]
161. Kim, S.W.; Lee, Y.K.; Kim, S.H.; Park, J.Y.; Lee, D.U.; Choi, J.; Hong, J.H.; Kim, S.; Khang, D. Covalent, Non-Covalent, Encapsulated Nanodrug Regulate the Fate of Intra- and Extracellular Trafficking: Impact on Cancer and Normal Cells. *Sci. Rep.* **2017**, *7*. [CrossRef]
162. Lahiani, M.H.; Eassa, S.; Parnell, C.; Nima, Z.; Ghosh, A.; Biris, A.S.; Khodakovskaya, M.V. Carbon nanotubes as carriers of Panax ginseng metabolites and enhancers of ginsenosides Rb1 and Rg1 anti-cancer activity. *Nanotechnology* **2017**, *28*. [CrossRef]
163. Wang, N.; Feng, Y.X.; Zeng, L.L.; Zhao, Z.N.; Chen, T.F. Functionalized Multiwalled Carbon Nanotubes as Carriers of Ruthenium Complexes to Antagonize Cancer Multidrug Resistance and Radioresistance. *ACS Appl. Mater. Inter.* **2015**, *7*, 14933–14945. [CrossRef]
164. Zakaria, A.; Picaud, F.; Rattier, T.; Pudlo, M.; Saviot, L.; Chassagnon, R.; Lherminier, J.; Gharbi, T.; Micheau, O.; Herlem, G. Nanovectorization of TRAIL with Single Wall Carbon Nanotubes Enhances Tumor Cell Killing. *Nano Lett.* **2015**, *15*, 891–895. [CrossRef]
165. Jiang, T.Y.; Sun, W.J.; Zhu, Q.W.; Burns, N.A.; Khan, S.A.; Mo, R.; Gu, Z. Furin-Mediated Sequential Delivery of Anticancer Cytokine and Small-Molecule Drug Shuttled by Graphene. *Adv. Mater.* **2015**, *27*, 1021–1028. [CrossRef] [PubMed]
166. Wang, S.B.; Ma, Y.Y.; Chen, X.Y.; Zhao, Y.Y.; Mou, X.Z. Ceramide-Graphene Oxide Nanoparticles Enhance Cytotoxicity and Decrease HCC Xenograft Development: A Novel Approach for Targeted Cancer Therapy. *Front. Pharmacol.* **2019**, *10*. [CrossRef] [PubMed]

167. Al Faraj, A.; Shaik, A.S.; Ratemi, E.; Halwani, R. Combination of drug-conjugated SWCNT nanocarriers for efficient therapy of cancer stem cells in a breast cancer animal model. *J. Control. Release* **2016**, *225*, 240–251. [CrossRef] [PubMed]
168. Jia, Y.J.; Weng, Z.Y.; Wang, C.Y.; Zhu, M.J.; Lu, Y.S.; Ding, L.L.; Wang, Y.K.; Cheng, X.H.; Lin, Q.; Wu, K.J. Increased chemosensitivity and radiosensitivity of human breast cancer cell lines treated with novel functionalized single-walled carbon nanotubes. *Oncol. Lett.* **2017**, *13*, 206–214. [CrossRef] [PubMed]
169. Luo, Y.A.; Cai, X.L.; Li, H.; Lin, Y.H.; Du, D. Hyaluronic Acid-Modified Multifunctional Q-Graphene for Targeted Killing of Drug-Resistant Lung Cancer Cells. *ACS Appl. Mater. Inter.* **2016**, *8*, 4048–4055. [CrossRef]
170. Nigam, P.; Waghmode, S.; Louis, M.; Wangnoo, S.; Chavan, P.; Sarkar, D. Graphene quantum dots conjugated albumin nanoparticles for targeted drug delivery and imaging of pancreatic cancer. *J. Mater. Chem. B* **2014**, *2*, 3190–3195. [CrossRef]
171. Vittorio, O.; Le Grand, M.; Makharza, S.A.; Curcio, M.; Tucci, P.; Iemma, F.; Nicoletta, F.P.; Hampel, S.; Cirillo, G. Doxorubicin synergism and resistance reversal in human neuroblastoma BE(2)C cell lines: An in vitro study with dextran-catechin nanohybrids. *Eur. J. Pharm. Biopharm.* **2018**, *122*, 176–185. [CrossRef]
172. Li, R.B.; Wu, R.; Zhao, L.; Wu, M.H.; Yang, L.; Zou, H.F. P-Glycoprotein Antibody Functionalized Carbon Nanotube Overcomes the Multidrug Resistance of Human Leukemia Cells. *ACS Nano* **2010**, *4*, 1399–1408. [CrossRef]
173. Zhang, H.J.; Xiong, J.; Guo, L.T.; Patel, N.; Guang, X.N. Integrated traditional Chinese and western medicine modulator for overcoming the multidrug resistance with carbon nanotubes. *RSC Adv.* **2015**, *5*, 71287–71296. [CrossRef]
174. Nowacki, M.; Wisniewski, M.; Werengowska-Ciecwierz, K.; Roszek, K.; Czarnecka, J.; Lakomska, I.; Kloskowski, T.; Tyloch, D.; Debski, R.; Pietkun, K.; et al. Nanovehicles as a novel target strategy for hyperthermic intraperitoneal chemotherapy: A multidisciplinary study of peritoneal carcinomatosis. *Oncotarget* **2015**, *6*, 22776–22798. [CrossRef]
175. Zhang, G.L.; Du, R.H.; Qian, J.C.; Zheng, X.J.; Tian, X.H.; Cai, D.Q.; He, J.C.; Wu, Y.Q.; Huang, W.; Wang, Y.Y.; et al. A tailored nanosheet decorated with a metallized dendrimer for angiography and magnetic resonance imaging-guided combined chemotherapy. *Nanoscale* **2018**, *10*, 488–498. [CrossRef]
176. Gu, Y.M.; Guo, Y.Z.; Wang, C.Y.; Xu, J.K.; Wu, J.P.; Kirk, T.B.; Ma, D.; Xue, W. A polyamidoamne dendrimer functionalized graphene oxide for DOX and MMP-9 shRNA plasmid co-delivery. *Mat. Sci. Eng. C* **2017**, *70*, 572–585. [CrossRef]
177. Cao, X.F.; Feng, F.L.; Wang, Y.S.; Yang, X.Y.; Duan, H.Q.; Chen, Y.S. Folic acid-conjugated graphene oxide as a transporter of chemotherapeutic drug and siRNA for reversal of cancer drug resistance. *J. Nanopart. Res.* **2013**, *15*. [CrossRef]
178. Liang, X.J.; Meng, H.; Wang, Y.Z.; He, H.Y.; Meng, J.; Lu, J.; Wang, P.C.; Zhao, Y.L.; Gao, X.Y.; Sun, B.Y.; et al. Metallofullerene nanoparticles circumvent tumor resistance to cisplatin by reactivating endocytosis. *Proc. Natl. Acad. Sci. USA* **2010**, *107*, 7449–7454. [CrossRef]
179. Tredan, O.; Galmarini, C.M.; Patel, K.; Tannock, I.F. Drug resistance and the solid tumor microenvironment. *J. Natl. Cancer Inst.* **2007**, *99*, 1441–1454. [CrossRef]
180. Rundqvist, H.; Johnson, R.S. Tumour oxygenation: Implications for breast cancer prognosis. *J. Intern. Med.* **2013**, *274*, 105–112. [CrossRef]
181. Vittorio, O.; Cojoc, M.; Curcio, M.; Spizzirri, U.G.; Hampel, S.; Nicoletta, F.P.; Iemma, F.; Dubrovska, A.; Kavallaris, M.; Cirillo, G. Polyphenol Conjugates by Immobilized Laccase: The Green Synthesis of Dextran-Catechin. *Macromol. Chem. Phys.* **2016**, *217*, 1488–1492. [CrossRef]
182. Septiadi, D.; Crippa, F.; Moore, T.L.; Rothen-Rutishauser, B.; Petri-Fink, A. Nanoparticle-Cell Interaction: A Cell Mechanics Perspective. *Adv. Mater.* **2018**, *30*. [CrossRef]
183. Rodriguez-Fernandez, L.; Valiente, R.; Gonzalez, J.; Villegas, J.C.; Fanarraga, M.L. Multiwalled Carbon Nanotubes Display Microtubule Biomimetic Properties in Vivo, Enhancing Microtubule Assembly and Stabilization. *ACS Nano* **2012**, *6*, 6614–6625. [CrossRef]
184. Madannejad, R.; Shoaie, N.; Jahanpeyma, F.; Darvishi, M.H.; Azimzadeh, M.; Javadi, H. Toxicity of carbon-based nanomaterials: Reviewing recent reports in medical and biological systems. *Chem.-Biol. Interact.* **2019**, *307*, 206–222. [CrossRef]
185. Yuan, X.; Zhang, X.X.; Sun, L.; Wei, Y.Q.; Wei, X.W. Cellular Toxicity and Immunological Effects of Carbon-based Nanomaterials. *Part. Fibre Toxicol.* **2019**, *16*. [CrossRef]

186. Li, Y.; Liu, Y.; Fu, Y.J.; Wei, T.T.; Le Guyader, L.; Gao, G.; Liu, R.S.; Chang, Y.Z.; Chen, C.Y. The triggering of apoptosis in macrophages by pristine graphene through the MAPK and TGF-beta signaling pathways. *Biomaterials* **2012**, *33*, 402–411. [CrossRef]
187. Gao, X.; Schottker, B. Reduction-oxidation pathways involved in cancer development: A systematic review of literature reviews. *Oncotarget* **2017**, *8*, 51888–51906. [CrossRef]
188. Garcia-Hevia, L.; Villegas, J.C.; Fernandez, F.; Casafont, I.; Gonzalez, J.; Valiente, R.; Fanarraga, M.L. Multiwalled Carbon Nanotubes Inhibit Tumor Progression in a Mouse Model. *Adv. Healthc. Mater.* **2016**, *5*, 1080–1087. [CrossRef]
189. González-Lavado, E.; Valdivia, L.; García-Castaño, A.; González, F.; Pesquera, C.; Valiente, R.; Fanarraga, M.L. Multi-walled carbon nanotubes complement the anti-tumoral effect of 5-Fluorouracil. *Oncotarget* **2019**, *10*, 2022–2029.
190. Ling, B.P.; Chen, H.T.; Liang, D.Y.; Lin, W.; Qi, X.Y.; Liu, H.P.; Deng, X. Acidic pH and High-H2O2 Dual Tumor Microenvironment-Responsive Nanocatalytic Graphene Oxide for Cancer Selective Therapy and Recognition. *ACS Appl. Mater. Inter.* **2019**, *11*, 11157–11166. [CrossRef]
191. Zhang, X.F.; Huang, F.H.; Zhang, G.L.; Bai, D.P.; Massimo, D.F.; Huang, Y.F.; Gurunathan, S. Novel biomolecule lycopene-reduced graphene oxide-silver nanoparticle enhances apoptotic potential of trichostatin A in human ovarian cancer cells (SKOV3). *Int. J. Nanomed.* **2017**, *12*, 7551–7575. [CrossRef]
192. Yang, C.; Peng, S.; Sun, Y.M.; Miao, H.T.; Lyu, M.; Ma, S.J.; Luo, Y.; Xiong, R.; Xie, C.H.; Quan, H. Development of a hypoxic nanocomposite containing high-Z element as 5-fluorouracil carrier activated self-amplified chemoradiotherapy co-enhancement. *Roy. Soc. Open Sci.* **2019**, *6*. [CrossRef]
193. Denkova, A.G.; de Kruijff, R.M.; Serra-Crespo, P. Nanocarrier-Mediated Photochemotherapy and Photoradiotherapy. *Adv. Healthc. Mater.* **2018**, *7*. [CrossRef]
194. Shanmugam, V.; Selvakumar, S.; Yeh, C.S. Near-infrared light-responsive nanomaterials in cancer therapeutics. *Chem. Soc. Rev.* **2014**, *43*, 6254–6287. [CrossRef]
195. MacDonald, I.J.; Dougherty, T.J. Basic principles of photodynamic therapy. *J. Porphyr. Phthalocyanines* **2001**, *5*, 105–129. [CrossRef]
196. Zhang, M.; Wang, W.T.; Wu, F.; Yuan, P.; Chi, C.; Zhou, N.L. Magnetic and fluorescent carbon nanotubes for dual modal imaging and photothermal and chemo-therapy of cancer cells in living mice. *Carbon* **2017**, *123*, 70–83. [CrossRef]
197. Chen, H.L.; Liu, Z.M.; Li, S.Y.; Su, C.K.; Qiu, X.J.; Zhong, H.Q.; Guo, Z.Y. Fabrication of Graphene and AuNP Core Polyaniline Shell Nanocomposites as Multifunctional Theranostic Platforms for SERS Real-time Monitoring and Chemo-photothermal Therapy. *Theranostics* **2016**, *6*, 1096–1104. [CrossRef] [PubMed]
198. Liu, J.Z.; Dong, J.; Zhang, T.; Peng, Q. Graphene-based nanomaterials and their potentials in advanced drug delivery and cancer therapy. *J. Control. Release* **2018**, *286*, 64–73. [CrossRef] [PubMed]
199. Fortner, J.D.; Lyon, D.Y.; Sayes, C.M.; Boyd, A.M.; Falkner, J.C.; Hotze, E.M.; Alemany, L.B.; Tao, Y.J.; Guo, W.; Ausman, K.D.; et al. C-60 in water: Nanocrystal formation and microbial response. *Environ. Sci. Technol.* **2005**, *39*, 4307–4316. [CrossRef] [PubMed]
200. Zhang, Q.; Yang, W.J.; Man, N.; Zheng, F.; Shen, Y.Y.; Sun, K.J.; Li, Y.; Wen, L.P. Autophagy-mediated chemosensitization in cancer cells by fullerene C60 nanocrystal. *Autophagy* **2009**, *5*, 1107–1117. [CrossRef]
201. Wei, P.F.; Zhang, L.; Lu, Y.; Man, N.; Wen, L.P. C60(Nd) nanoparticles enhance chemotherapeutic susceptibility of cancer cells by modulation of autophagy. *Nanotechnology* **2010**, *21*. [CrossRef]
202. Mocan, T.; Matea, C.T.; Cojocaru, I.; Ilie, I.; Tabaran, F.A.; Zaharie, F.; Iancu, C.; Bartos, D.; Mocan, L. Photothermal Treatment of Human Pancreatic Cancer Using PEGylated Multi-Walled Carbon Nanotubes Induces Apoptosis by Triggering Mitochondrial Membrane Depolarization Mechanism. *J. Cancer* **2014**, *5*, 679–688. [CrossRef]
203. Burke, A.R.; Singh, R.N.; Carroll, D.L.; Wood, J.C.S.; D'Agostino, R.B.; Ajayan, P.M.; Torti, F.M.; Torti, S.V. The resistance of breast cancer stem cells to conventional hyperthermia and their sensitivity to nanoparticle-mediated photothermal therapy. *Biomaterials* **2012**, *33*, 2961–2970. [CrossRef]
204. Suo, X.B.; Eldridge, B.N.; Zhang, H.; Mao, C.Q.; Min, Y.Z.; Sun, Y.; Singh, R.; Ming, X. P-Glycoprotein-Targeted Photothermal Therapy of Drug-Resistant Cancer Cells Using Antibody-Conjugated Carbon Nanotubes. *ACS Appl. Mater. Inter.* **2018**, *10*, 33464–33473. [CrossRef]

205. Wang, J.L.; Wang, R.; Zhang, F.R.; Yin, Y.J.; Mei, L.X.; Song, F.J.; Tao, M.T.; Yue, W.Q.; Zhong, W.Y. Overcoming multidrug resistance by a combination of chemotherapy and photothermal therapy mediated by carbon nanohorns. *J. Mater. Chem. B* **2016**, *4*, 6043–6051. [CrossRef] [PubMed]
206. Bhirde, A.A.; Chikkaveeraiah, B.V.; Srivatsan, A.; Niu, G.; Jin, A.J.; Kapoor, A.; Wang, Z.; Patel, S.; Patel, V.; Gorbach, A.M.; et al. Targeted Therapeutic Nanotubes Influence the Viscoelasticity of Cancer Cells to Overcome Drug Resistance. *ACS Nano* **2014**, *8*, 4177–4189. [CrossRef] [PubMed]
207. Chen, A.P.; Xu, C.; Li, M.; Zhang, H.L.; Wang, D.C.; Xia, M.; Meng, G.; Kang, B.; Chen, H.Y.; Wei, J.W. Photoacoustic "nanobombs" fight against undesirable vesicular compartmentalization of anticancer drugs. *Sci. Rep.* **2015**, *5*. [CrossRef] [PubMed]
208. Wang, C.H.; Chiou, S.H.; Chou, C.P.; Chen, Y.C.; Huang, Y.J.; Peng, C.A. Photothermolysis of glioblastoma stem-like cells targeted by carbon nanotubes conjugated with CD133 monoclonal antibody. *Nanomedicine* **2011**, *7*, 69–79. [CrossRef]
209. Zhou, F.F.; Wu, S.; Song, S.; Chen, W.R.; Resasco, D.E.; Xing, D. Antitumor immunologically modified carbon nanotubes for photothermal therapy. *Biomaterials* **2012**, *33*, 3235–3242. [CrossRef]
210. Thapa, R.K.; Byeon, J.H.; Choi, H.G.; Yong, C.S.; Kim, J.O. PEGylated lipid bilayer-wrapped nanographene oxides for synergistic co-delivery of doxorubicin and rapamycin to prevent drug resistance in cancers. *Nanotechnology* **2017**, *28*. [CrossRef]
211. Tran, T.H.; Nguyen, H.T.; Pham, T.T.; Choi, J.Y.; Choi, H.G.; Yong, C.S.; Kim, J.O. Development of a Graphene Oxide Nanocarrier for Dual-Drug Chemo-photothermy to Overcome Drug Resistance in Cancer. *ACS Appl. Mater. Inter.* **2015**, *7*, 28647–28655. [CrossRef]
212. Yuan, Y.; Zhang, Y.Q.; Liu, B.; Wu, H.M.; Kang, Y.J.; Li, M.; Zeng, X.; He, N.Y.; Zhang, G. The effects of multifunctional MiR-122-loaded graphene-gold composites on drug-resistant liver cancer. *J. Nanobiotechnol.* **2015**, *13*. [CrossRef]
213. Hou, L.; Feng, Q.H.; Wang, Y.T.; Yang, X.M.; Ren, J.X.; Shi, Y.Y.; Shan, X.N.; Yuan, Y.J.; Wang, Y.C.; Zhang, Z.Z. Multifunctional hyaluronic acid modified graphene oxide loaded with mitoxantrone for overcoming drug resistance in cancer. *Nanotechnology* **2016**, *27*. [CrossRef]
214. Wang, M.; Xiao, Y.; Li, Y.; Wu, J.H.; Li, F.Y.; Ling, D.S.; Gao, J.Q. Reactive oxygen species and near-infrared light dual-responsive indocyanine green-loaded nanohybrids for overcoming tumour multidrug resistance. *Eur. J. Pharm. Sci.* **2019**, *134*, 185–193. [CrossRef]
215. Wang, M.; Wu, J.H.; Li, Y.; Li, F.Y.; Hu, X.; Wang, G.; Han, M.; Ling, D.S.; Gao, J.Q. A tumor targeted near-infrared light-controlled nanocomposite to combat with multidrug resistance of cancer. *J. Control. Release* **2018**, *288*, 34–44. [CrossRef] [PubMed]
216. Feng, L.Z.; Li, K.Y.; Shi, X.Z.; Gao, M.; Liu, J.; Liu, Z. Smart pH-Responsive Nanocarriers Based on Nano-Graphene Oxide for Combined Chemo- and Photothermal Therapy Overcoming Drug Resistance. *Adv. Healthc. Mater.* **2014**, *3*, 1261–1271. [CrossRef] [PubMed]
217. Zeng, Y.P.; Yang, Z.Y.; Li, H.; Hao, Y.H.; Liu, C.; Zhu, L.; Liu, J.; Lu, B.H.; Li, R. Multifunctional Nanographene Oxide for Targeted Gene-Mediated Thermochemotherapy of Drug-resistant Tumour. *Sci. Rep.* **2017**, *7*. [CrossRef]
218. Zhao, Y.; Zhang, C.R.; Gao, L.Q.; Yu, X.H.; Lai, J.H.; Lu, D.H.; Bao, R.; Wang, Y.P.; Jia, B.; Wang, F.; et al. Chemotherapy-Induced Macrophage Infiltration into Tumors Enhances Nanographene-Based Photodynamic Therapy. *Cancer Res.* **2017**, *77*, 6021–6032. [CrossRef] [PubMed]
219. Sasidharan, A.; Sivaram, A.J.; Retnakumari, A.P.; Chandran, P.; Malarvizhi, G.L.; Nair, S.; Koyakutty, M. Radiofrequency Ablation of Drug-Resistant Cancer Cells Using Molecularly Targeted Carboxyl-Functionalized Biodegradable Graphene. *Adv. Healthc. Mater.* **2015**, *4*, 679–684. [CrossRef]
220. Chen, Y.; Xu, P.F.; Shu, Z.; Wu, M.Y.; Wang, L.Z.; Zhang, S.J.; Zheng, Y.Y.; Chen, H.R.; Wang, J.; Li, Y.P.; et al. Multifunctional Graphene Oxide-based Triple Stimuli-Responsive Nanotheranostics. *Adv. Funct. Mater.* **2014**, *24*, 4386–4396. [CrossRef]

© 2020 by the authors. Licensee MDPI, Basel, Switzerland. This article is an open access article distributed under the terms and conditions of the Creative Commons Attribution (CC BY) license (http://creativecommons.org/licenses/by/4.0/).

Article

Graphene Aerogel Growth on Functionalized Carbon Fibers

Katerina Vrettos [1], Konstantinos Spyrou [2] and Vasilios Georgakilas [1,*]

[1] Department of Material Science, University of Patras, 26504 Patras, Greece; c.vrettos0@gmail.com
[2] Department of Materials Science and Engineering, University of Ioannina, 45110 Ioannina, Greece; konstantinos.spyrou1@gmail.com
* Correspondence: viegeorgaki@upatras.gr; Tel.: +30-2610-996321

Academic Editors: Long Y. Chiang and Giuseppe Cirillo
Received: 28 January 2020; Accepted: 9 March 2020; Published: 12 March 2020

Abstract: Graphene aerogel (GA) is a lightweight, porous, environmentally friendly, 3D structured material with interesting properties, such as electrical conductivity, a high surface area, and chemical stability, which make it a powerful tool in energy storage, sensing, catalyst support, or environmental applications. However, the poor mechanical stability that often characterizes graphene aerogels is a serious obstacle for their use in such applications. Therefore, we report here the successful mechanical reinforcement of GA with carbon fibers (CFs) by combining reduced graphene oxide (rGO) and CFs in a composite material. The surfaces of the CFs were first successfully desized and enriched with epoxy groups using epichloridrine. Epoxy-functionalized CFs (epoxy-CFs) were further covered by reduced graphene oxide (rGO) nanosheets, using triethylene tetramine (TETA) as a linker. The rGO-covered CFs were finally incorporated into the GA, affording a stiff monolithic aerogel composite. The as-prepared epoxy-CF-reinforced GA was characterized by spectroscopic and microscopic techniques and showed enhanced electrical conductivity and compressive strength. The improved electrical and mechanical properties of the GA-CFs composite could be used, among other things, as electrode material or strain sensor applications.

Keywords: carbon fibers; surface treatment; grafting; graphene aerogel

1. Introduction

Graphene oxide (GO) is the most common graphene derivative [1]. It is formed by the oxidative treatment of graphite and is characterized by a large amount of oxygen groups, which are spread over the graphenic surface. Carboxylates mainly at the edges, epoxy, and hydroxylates at the core provide a strong hydrophilic character and dramatically reduce the aromatic character of GO. The treatment of GO with reducing agents leads to the partial removal of the oxygen groups and the reconstruction of aromaticity, to the so-called reduced GO (rGO). Depending on the reductive treatment, rGO shows often remarkable electrical conductivity, comparable with pristine graphene. On the other hand, the removal of the majority of the oxygen groups results in the decrease in the hydrophilic character of rGO and the formation of aggregates when they are dispersed in water [2].

Hydrothermal reductive treatment of GO often leads to the formation of a stable rGO hydrogel, depending on the conditions. GO is a highly hydrophilic 2D material, which, under reductive conditions, is partially gaining its aromaticity and self-assembled in a 3D structure due to π-π stacking interactions. The oxygen groups on the graphene surface, as well as the entrapped hydrophilic reducing agent, are often responsible for the entrapment of a large amount water between the rGO nanosheets, leading to the formation of stable hydrogels, which can be transformed into a graphene aerogel (GA), after water removal by a freeze-drying procedure.

GA is a lightweight, porous, environmentally friendly 3D structured material, with electrical conductivity, chemical inertness and a high surface area [3–6]. It could be used in several applications, including in supercapacitors, lithium ion batteries, fuel or solar cells, and for environmental purposes such as water purification, gas separation, or electromagnetic interference shielding. The use of GA in most applications could be largely promoted by a significant mechanical reinforcement, which is a real challenge, taking into consideration the poor mechanical properties of GA [3–6].

On the other hand, carbon fibers (CFs) are a graphitic material that have been used widely for the reinforcement of polymers, due to their remarkable mechanical, thermal, and electrical properties as well. CFs/polymer composites have been used in applications such as in aerospace, automotive, nuclear engineering, where strength, stiffness, and lightweight materials are critical requirements [7–11]. However, their smooth and inert graphitic surface, almost free of functional groups, does not favor the adhesion with matrix polymer molecules, weakening the load that is transferred from the matrix to the CFs, and limiting the mechanical reinforcement [8,9]. Therefore, in the past decades, many research efforts have been made to modify chemically the surface of CFs, and thus to improve the interfacial adhesion in the resulting composites. However, there has been limited success up to now in the sufficient functionalization of the CF's surface. In most cases, the CF's surface is treated with strong acids, such as a nitric or nitric/sulfuric acid mixture, to introduce a limited amount of oxygen groups (mainly carboxylates) [12–14]. Carboxylates are then used as reactive sites to graft on the CF's surface coupling agents such as simple diamines, dialdehydes [15,16], or, more specifically, polyphosphazenes and siloxanes [17–19]. The decoration of CFs with coupling agents increases the wettability and chemical reactivity of CFs, and this improves the adhesion between the CFs and polymer matrices. Recently, GO has been also used to modify the CF's surface, improving the interfacial properties of the composite material [20]. Furthermore, efforts have appeared in the literature that combined CFs and GA in a 3D composite structure, ideally having the properties of both components [21,22].

In the present article, we describe an effective enrichment of CFs surface with epoxy groups, and their successful incorporation in a GA, forming a 3D structured monolithic aerogel composite with improved electrical and mechanical properties. The CFs were first desized and then functionalized with epoxy groups to enhance the binding sites of the CFs, using, for the first time, epichlorohydrin. Epoxy-functionalized CFs (epoxy-CFs) were then covered by rGO nanosheets and finally incorporated successfully into the GA, during a sol-gel hydrothermal reduction process of GO. The as-prepared CFs/rGO monolithic aerogel composite was fully characterized by spectroscopic and microscopic techniques, and finally showed enhanced electrical conductivity and compression stability. Due to these characteristics, the CFs/rGO aerogel composite could be used, among other things, as an electrode material or in strain sensor applications [23].

2. Results and Discussion

Initially, the CFs were desized in acetone [24] and then oxidized using an acidic treatment under ultrasonication. After this procedure, a plethora of active oxygen groups, mainly hydroxylates and carboxylates, appeared on the surface of the CFs, increasing the total surface energy and polarity, which was helpful to improve the wettability of the CFs [25,26]. The morphology of the CFs was studied using scanning electron microscopy (SEM), and characteristic micrographs are given in Figure 1.

The polymer removal alters the morphology and increases the roughness of the CFs' surface. The characteristic change in the roughness of CFs after desizing is indicated by atomic force microscopy (AFM) images, as shown in Figure 2. The surfaces of untreated CFs seem to be relatively neat and smooth. Few narrow grooves, distributed in parallel along the longitudinal direction of the fiber, are due to the fiber manufacture process. Compared with the untreated CFs (Figure 1a), the fiber surface becomes rougher after oxidative treatment.

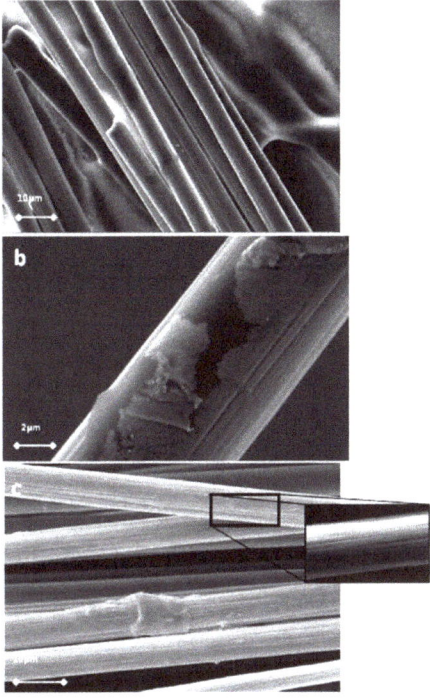

Figure 1. Scanning electron microscopy (SEM) images of carbon fibers (CFs) before (**a**) and after the treatment with acetone (**b**,**c**).

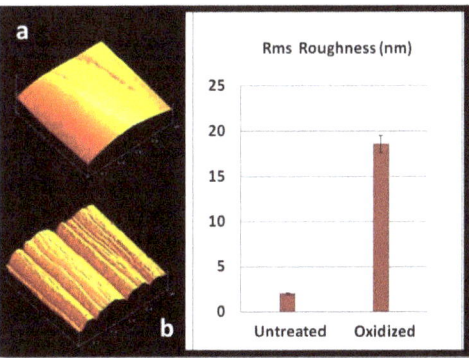

Figure 2. Atomic force microscopy (AFM) topography images of (**a**) untreated and (**b**) oxidized CFs.

Thermogravimetric and differential thermal analysis (TG-DTA) curves of pristine CFs and oxidized CFs are depicted in Figure 3. Pristine CFs were stable until 700 °C in the air, while, after desizing and acid treatment, oxidized CFs were decomposed much easier between 500 and 700 °C (see Figure 3). The decomposition of both the pristine and oxidized CFs was accompanied by an analogous exothermic peak, recorded in DTA as expected. Pristine CFs also indicated a 2% weight loss between 350 and 500 °C, which could be attributed to the removal of the sizing agent.

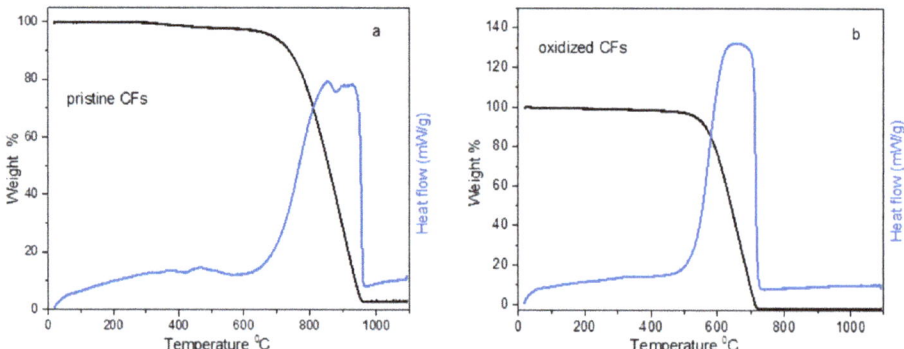

Figure 3. Thermo gravimetric analysis (TGA) and heat flow graphs of pristine (**a**) and oxidized CFs (**b**).

The Fourier–transform infrared spectrometer (FT-IR) spectrum of the oxidized CFs (see line b in Figure 4) showed a few weak but characteristic peaks due to the presence of oxygen groups on their surface, in contrast with the featureless spectrum of pristine CFs (see line a in Figure 4). The peaks at 3300 (OH stretching vibrations) and 1033 cm^{-1} (C-O stretching vibrations) indicated the appearance of hydroxyl groups on the CF surface. The peaks at 1715 and 1640 cm^{-1} (C=O stretching) could be attributed to the presence of carboxylates and carbonyl groups, respectively, while peaks at 2860, 2930, and 1335 cm^{-1} indicate sp^3 C-H stretching [27–29].

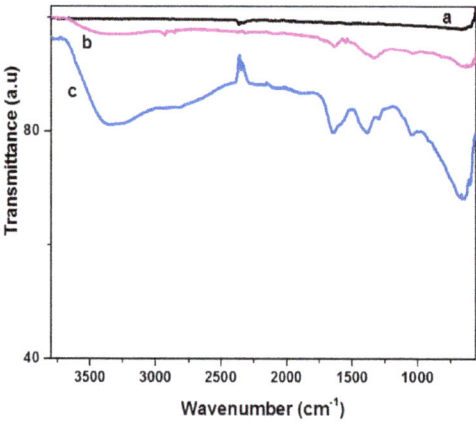

Figure 4. Fourier–transform infrared spectrometer (FT-IR) spectrum of (**a**) pristine CFs (**b**) acid treated CFs and (**c**) epoxy-functionalized CFs (epoxy-CFs).

Epichlorohydrin was then grafted on the CFs following mainly two different pathways. It can be added to the carboxylates or OH groups using the epoxy or chlorine end, respectively, as shown in Figure 5. Both pathways in the alkaline environment lead to the formation of epoxy groups on the surface of the CFs (epoxy-CFs). The FT-IR spectrum of the epoxy-CFs showed their enrichment with hydroxyl and epoxy groups, as indicated by the OH stretching vibration at 3330 cm^{-1} and C-O-C, epoxy stretching vibration at 1290 and 1050 cm^{-1}. The characteristic CH and C=C vibrations at 2850 (stretching), 1400 (in plane bending), and 1647 cm^{-1} (stretching), respectively, also appeared (see line c in Figure 4).

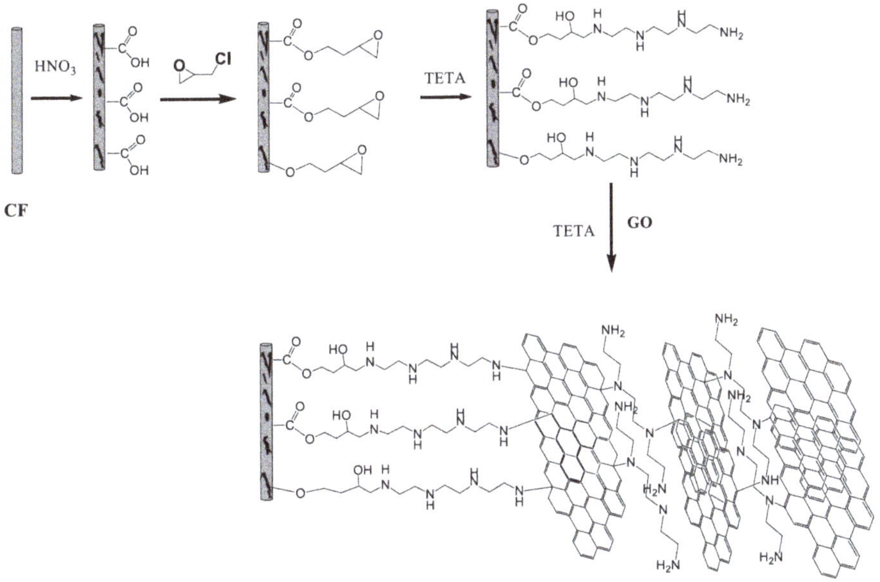

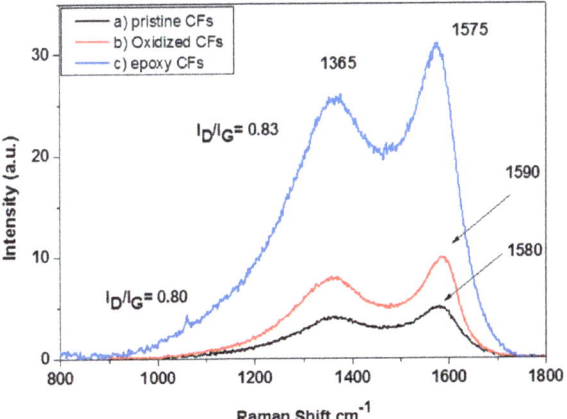

Figure 5. Schematic representation of the formation of CFs/ reduced GO (rGO) composite aerogel.

The Raman spectra of pristine, oxidized and epoxy-CFs are shown in Figure 6. The spectrum of the CFs contains two characteristic peaks, which are assigned to the graphitic E_{2g} G mode at ~1580 cm^{-1} and the disorder D mode at ~1365 cm^{-1}. In the spectrum of epoxy-CFs, the G band appeared slightly shifted to 1590 cm^{-1} due to the contribution of the D' band at around 1610 cm^{-1}, which is more intense after the introduction of oxygen groups on the CFs' surface. The I_D/I_G ratio after the reaction with epichlorohydrin was slightly increased due to the introduction of epoxy groups [30].

Figure 6. Raman spectra of (**a**) pristine (**b**) oxidized and (**c**) epoxy-CFs.

In order to identify further the functional chemical groups of the epoxy-CFs, X-ray photoelectron spectroscopy (XPS) measurements were employed. From the C1s high resolution photoelectron spectra (Figure 7), several changes after the chemical modification were deduced. The most important information that was collected here was the reduction in the C-C frame from 80.3% for the pristine

CFs to 29.6% for the epoxy-CFs. The peak at 286.1 eV is increased because of the addition of C-O and also C-N bonds. The existence of these two functionalities explains the small shift from 286.1 eV, due to CFs, to 285.8 eV. Important evidence for the successful functionalization of CFs is the significant increase in the photoelectron peak at 286.8 eV from 3.1% to 23.0%. This increase is due to the epoxy group created after the chemical functionalization [29,31].

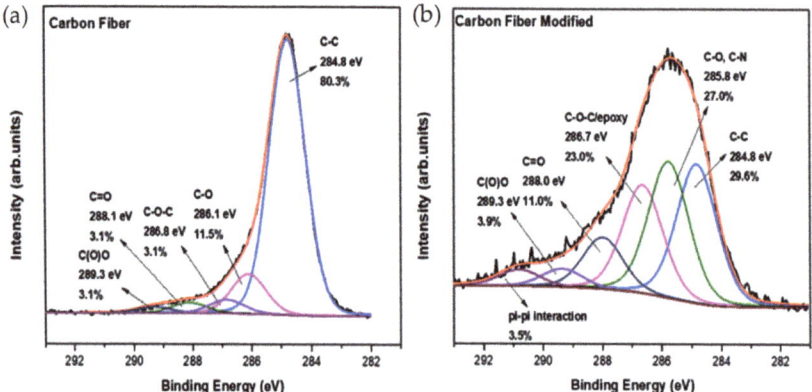

Figure 7. C1s photoelectron peak of (**a**) pristine CFs and (**b**) epoxy-CFs.

Epoxy-CFs were transferred to a diluted GO dispersion in an alkaline solution of triethylene tetramine (TETA) and heated hydrothermally at 100 °C in a sealed bottle. At this stage, the brown GO solution becomes colorless after the reaction, indicating the successful immobilization of the rGO nanosheets on the external surface of the CFs (see Figure 8a,b). The rGO-covered epoxy-CFs finally underwent a second hydrothermal treatment with a larger amount of GO dispersed in alkaline solution of TETA. Under these conditions, the GO was reduced partially and aggregated, forming a stable hydrogel (see Figure 8c). Through the nucleophilic addition to the epoxy ring opening functionalization, TETA acts here as a reducing and coupling agent on both CFs and rGO nanosheets, and thus contributes significantly to the hydrogel formation as a bridge molecule. The insertion of epoxy-CFs between the rGO nanosheets resulted in the successful incorporation of the former in the hydrogel (see Figure 8c). Finally, after freeze drying, the CFs-supported GA (CFs/rGO aerogel) was formed (see Figure 8d).

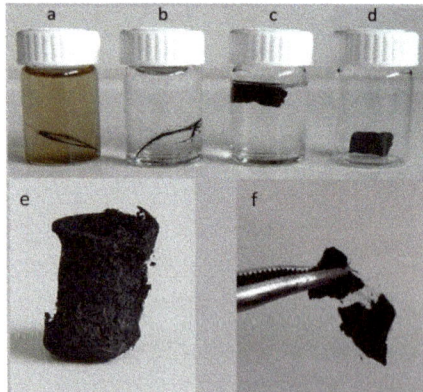

Figure 8. Photo of (**a**) epoxy-CFs in an alkaline solution of GO, (**b**) epoxy-CFs covered by the rGO nanosheets after hydrothermal heating, (**c**) CFs/rGO hydrogel and (**d,e**) aerogel, (**f**) CFs/rGO aerogel separated in two pieces using forcepts, where CFs are revealed from the internal.

CFs have a mean size of 0.3–0.5 cm and were randomly oriented in the GA, as observed optically under a microscope. This is a consequence of the random dispersion of CFs into the mixture of the GO before the hydrothermal treatment. CFs/rGO aerogel can be also formed in a single stage hydrothermal procedure by directly dispersing epoxy-CFs in a concentrated dispersion of GO in alkaline solution of TETA. In Figure 9, several characteristic images of rGO-covered epoxy-CFs are presented (Figure 9a,b,f), in comparison with poorly rGO-covered pristine CFs (Figure 9c,d) that were formed when pristine CFs were used instead of epoxy-CFs. As shown in Figure 9a–d, rGO nanosheets have extensively covered the surface of the epoxy-CFs, while the absence of epoxy groups in pristine CFs leads to much less coverage by rGO, as shown in Figure 9c,d. Finally, Figure 9e,f indicate the successful incorporation of CFs in the CFs/rGO aerogel. Due to the interaction between the epoxy-CFs and GO in the presence of the TETA bridge molecules, the epoxy-CFs reinforced GA were more condensed, having a lower volume (0.6 cm^3) than the pure rGO aerogel (0.9 cm^3). In addition, taking into consideration the masses of the components and the final products, epoxy-CFs reinforced GA showed a higher density (31.6 mg/cm^3) in comparison to pure rGO aerogel (12,5 mg/cm^3), due to the lower volume and the presence of CFs. The mass fraction of CFs in the CFs/rGO aerogel was estimated to be 0.31 and the volume fraction to be 5×10^{-3}.

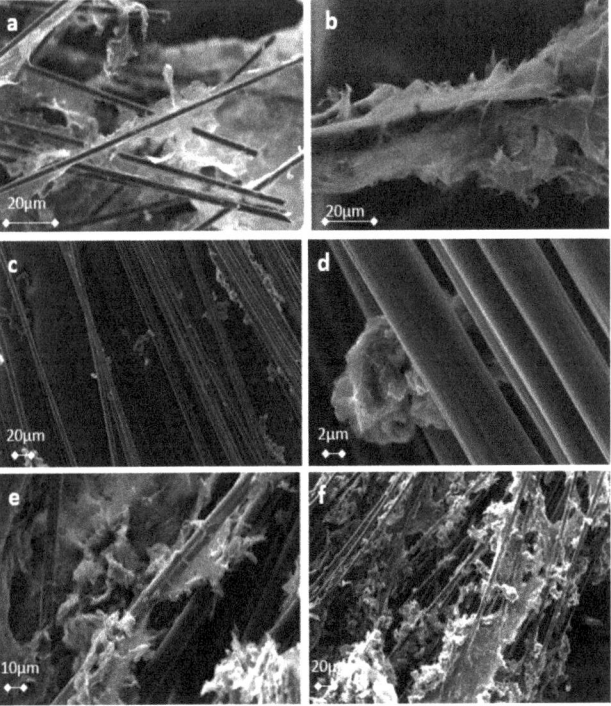

Figure 9. SEM images of epoxy-CFs (**a**,**b**) and pristine CFs (**c**,**d**) covered by rGO nanosheets, and CFs/rGO aerogel (**e**,**f**).

2.1. Electrical Conductivity

It is known that GAs are electrically conductive, due to the recovered aromaticity after the reduction of GO. CFs are also conductive, and thus the final CFs/rGO composite is highly conductive as expected (see Figure 10). In fact, the resistivity of a CFs/rGO aerogel monolith was measured to 28.8 Ω m, while an analogous rGO aerogel monolith was measured to 129.6 Ω m (see Table 1). The samples had a cylindrical shape and the resistance was measured by adapting two electrodes at the upper and lower surface of the cylinders (see experimental part). It is important to note here that the orientation of the CFs in the aerogel was random and not involved with the increased conductivity of the epoxy-CFs supported GA.

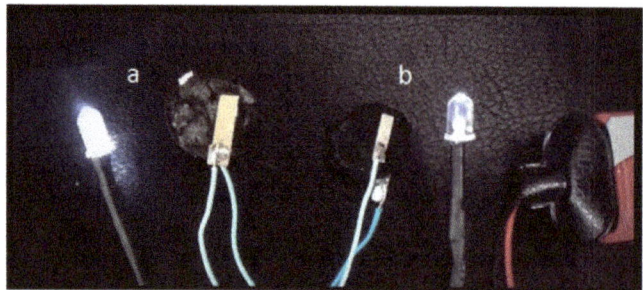

Figure 10. Electrical circuits with (**a**) CFs/rGO and (**b**) rGO aerogel monoliths. The intense light of the LED lamp in circuit (**a**) indicates the increased conductivity of CFs/rGO aerogel.

Table 1. Bulk resistance (R) and resistivity (r) of rGO and CFs/rGO aerogels.

	$S\ (m^2) \times 10^{-4}$	l (m)	R (kΩ)	ϱ (Ω m)
rGO	1.8	0.005	3.6	129.6
CFs/rGO	1.8	0.005	0.8	28.8

2.2. Mechanical Reinforcement

In a recent previous article, we showed that TETA-promoted rGO aerogels can be compressed to about 50% of the initial thickness by the placement of a 50 g standard weight on a GA cylindrical monolith, while the rGO aerogels promoted by aromatic diamines were compressed almost elastically [32]. Here, we demonstrate that TETA-promoted GA reinforced with CFs was not compressed under the same conditions, indicating the remarkable role of CFs on the mechanical reinforcement of the CFs/rGO aerogel (see Figure 11a–c). In contrast, GA reinforced with pristine CFs by the same procedure is fragile and mechanically very unstable, leading to negative results as regards mechanical measurements. This fact indicated the crucial role of epoxy groups in the successful incorporation of CFs in the GA.

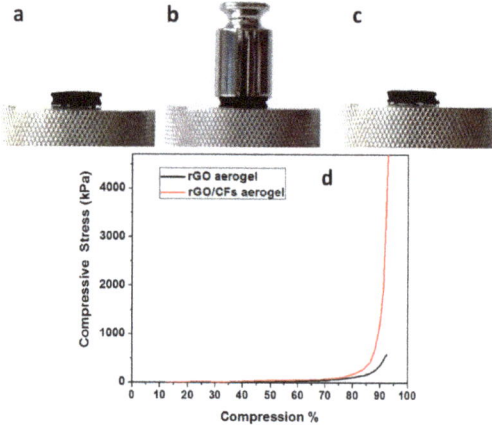

Figure 11. (Upper) CFs/rGO cylindrical aerogel monolith before (**a**) during (**b**) and after (**c**) the compression, with a 50 g standard weight. (**d**) The stress/compression diagram of rGO and CFs/rGO aerogel monoliths.

A similar conclusion was drawn by comparing the diagrams of compressive stress to compression of the rGO and the CFs/rGO aerogels (see Figure 11d). In fact, the rGO aerogel was compressed to about 90% with a 400 kPa stress, while in the case of the CFs/rGO aerogel, a remarkably higher compressive stress—about 1300 kPa—was applied to achieve a similar compression. This 3D porous, conductive, and highly stable CFs/rGO structure could become a highly promising material for applications in lightweight conductive cables, energy storage, catalysts, and functional textiles. The unique structure of those materials paves the way to design and fabricate lightweight porous materials with high performance.

3. Materials and Methods

3.1. Materials

Commercially available CFs, T700SC (Fiber Max, Volos, Greece) were used in this work. Epichlorohydrin (Alfa Aesar, Kandel, Germany), sodium hydroxide (NaOH, Sigma-Aldrich, St Louis city, MO, USA), ammonium hydroxide (NH_3, CARLO ERBA Reagents S.A.S., Barcelona, Spain), TETA (Sigma-Aldrich, St Louis city, MO, USA), nitric acid (HNO_3, 65%, CARLO ERBA Reagents S.A.S., Barcelona, Spain), and powder graphite (Sigma-Aldrich, St Louis city, MO, USA) were used without further treatment. GO was prepared in the lab, according to Staudenmaier's method [33] and the synthesis is described in detail, in ref [34].

3.2. Characterization

TGA/DSC measurements were performed on pristine and oxidized CFs (~10 mg) by means of a SETARAM SETSYS Evolution 18 Analyzer (SETARAM Instrumentation, Caluire, France) with Al_2O_3 crucibles, in the range of 25–1100 °C. A heating rate of 10 °C/min under air flow (16 mL/min) while used and purging was applied well before initiating the heating ramp. Buoyancy corrections were carried out through blank measurements.

Scanning electron microscopy (SEM) was carried out on a Zeiss EVO-MA10 (Carl Zeiss Microscopy GmbH, Jena, Germany). Infrared spectra were measured on a Fourier–transform infrared spectrometer (FT-IR) using the ATR technique on an IRTracer-100 Shimadzu spectrometer (Shimadzu Europa GmbH, Duisburg, Germany). Raman spectra were collected with a Raman system Lab-Ram HR Evolution RM (Horiba-Scientific, Kyoto, Japan) using a laser excitation line at 532 nm (laser diode). The laser power was 1.082 mV. All Raman parameters have been carefully controlled to avoid changes in the graphene materials. Bulk resistance was measured using a Keithley 2401 multimeter (Keithley Instruments, Solon, OH, USA), using two indium tin oxide (ITO) glass slides as electrodes that covered the upper and lower surface of the samples. Atomic force microscopy (AFM) measurements were performed in tapping mode with a multimode Nanoscope IIIa (Bruker, Billerica, MA, USA), using RTESPA-300 silicon cantilevers with a nominal tip radius 8 nm. The values of the stress/compression diagrams of rGO and rGO/CFs aerogel monoliths were recorded with a Hounsfield H20K-W test machine (rate 1.5 mm/min, Hounsfield Test Equipment, Red Hill, England).

X-ray photoelectron spectroscopy (XPS) measurements were performed in an ultra-high vacuum at a base pressure of 2×10^{-10} mbar, with a SPECS GmbH spectrometer (SPECS Surface Nano Analysis GmbH, Berlin, Germany) equipped with a monochromatic MgKa source (hv = 1253.6 eV), and a Phoibos-100 hemispherical analyzer (SPECS Surface Nano Analysis GmbH, Berlin, Germany). The spectra were collected in normal emission and the energy resolution was set to 1.16 eV to minimize measuring time. All binding energies were referenced to the C1s core level at 284.8 eV. Spectral analysis included a Shirley background subtraction and a peak deconvolution employing mixed Gaussian-Lorentzian functions, in a least square curve-fitting program (WinSpec) developed at the Laboratoire Interdisciplinaire de Spectroscopie Electronique, University of Namur, Belgium.

3.3. Oxidation Treatment and Functionalization of CFs

Some 100 mg of CFs were heated in acetone for 48 h at 60 °C. After the drying of the desized CFs, they were oxidized in conc. HNO_3 at 100 °C for 2 h in a sonication bath. Subsequently, the oxidized CFs were washed several times with deionized water until reaching pH ~ 7 and dried at 100 °C under vacuum. Oxidized CFs were placed in a solution of epichlorohydrin, while an ethanolic solution of NaOH was added slowly during refluxing (95 °C) for 3 h. The epoxy-CFs were finally washed with acetone and dried under a vacuum.

3.4. Formation of CFs/rGO Aerogel

Epoxy-CFs (6 mg) were placed into a concentrated solution of TETA in water (50% v/v), at 80 °C for 24 h. After a thorough washing, the as-prepared amine functionalized CFs (amino-CFs) were transferred in an alkaline GO dispersion (2 mg of GO, 50 μL conc. NH_3 in 20 mL H_2O) at 95 °C for 24 h. The as prepared rGO functionalized CFs were then placed in a dispersion of GO (10 mg) and TETA (10 μL) in 20 mL of water, and the mixture was heated in a sealed bottle at 95 °C for 24 h. The resulting hydrogel was washed several times with water and lyophilized for 24 h.

4. Conclusions

In this work, we demonstrated that epoxy-functionalized CFs can be successfully incorporated into GA by grafting rGO to their surface and forming a composite aerogel monolith with improved electrical and mechanical properties, due to the presence of CFs. The role of epichlorohydrin in the introduction of epoxy groups to the CFs surface was crucial, since epoxy groups are the key for the successful incorporation of CFs into rGO aerogel. The as-prepared CFs/rGO aerogel showed at least four times lower electrical resistivity than rGO aerogel, since the desized CFs function as conducting pathways within the porous structure. Despite desizing, the contribution of chemical functionalization to the surface of CFs to the mechanical properties of the final composite was also remarkable.

Author Contributions: Conceptualization, V.G. and K.V., methodology K.V., K.S., writing—original draft preparation, V.G. and K.V., K.S., writing—review and editing V.G. and K.V., supervision, V.G. All authors have read and agreed to the published version of the manuscript.

Funding: This research received no external funding.

Acknowledgments: This research is co-financed by Greece and the European Union (European Social Fund- ESF) through the Operational Programme «Human Resources Development, Education and Lifelong Learning» in the context of the project "Strengthening Human Resources Research Potential via Doctorate Research" (MIS-5000432), implemented by the State Scholarships Foundation (IKY).

Conflicts of Interest: The authors declare no conflicts of interest.

References and Note

1. Chen, D.; Feng, H.; Li, J. Graphene Oxide: Preparation, Functionalization, and Electrochemical Applications Information. *Chem. Rev.* **2012**, *112*, 6027–6053. [CrossRef] [PubMed]
2. Pei, S.; Cheng, H.M. The reduction of graphene oxide. *Carbon* **2012**, *50*, 3210–3228. [CrossRef]
3. Hu, H.; Zhao, Z.; Wan, W.; Gogotsi, Y.; Qiu, J. Ultralight and Highly Compressible Graphene Aerogels. *Adv. Mater.* **2013**, *25*, 2219–2223. [CrossRef] [PubMed]
4. Tang, G.; Jiang, Z.G.; Li, X.; Zhang, H.B.; Dasari, A.; Yu, Z.Z. Three-dimensional graphene aerogels and their electrically conductive composites. *Carbon* **2014**, *77*, 592–599. [CrossRef]
5. Mao, J.; Iocozzia, J.; Huang, J.; Meng, K.; Lai, Y.; Lin, Z. Graphene aerogels for efficient energy storage and conversion. *Energy Environ. Sci.* **2018**, *11*, 772–799. [CrossRef]
6. Li, C.; Ding, M.; Zhang, B.; Qiao, X.; Liu, C.Y. Graphene aerogels that withstand extreme compressive stress and strain. *Nanoscale* **2018**, *10*, 18291–18299. [CrossRef]
7. Prashanth, S.; Subbaya, K.M.; Nithin, K.; Sachhidananda, S. Fiber Reinforced Composites-A Review. *J. Mater. Sci. Eng.* **2017**, *6*, 341. [CrossRef]

8. Chen, J.; Xu, H.; Liu, C.; Mi, L.; Shen, C. The effect of double grafted interface layer on the properties of carbon fiber reinforced polyamide 66 composites. *Compos. Sci. Technol.* **2018**, *168*, 20–27. [CrossRef]
9. Koutroumanis, N.; Manikas, A.C.; Pappas, P.N.; Petropoulos, F.; Sygellou, L.; Tasis, D.; Papagelis, K.; Anagnostopoulos, G.; Galiotis, C. A novel mild method for surface treatment of carbon fibres in epoxy matrix Composites. *Compos. Sci. Technol.* **2018**, *157*, 178–184. [CrossRef]
10. Rajak, D.K.; Pagar, D.D.; Menezes, P.L.; Linul, E. Fiber-Reinforced Polymer Composites: Manufacturing, Properties, and Applications. *Polymers* **2019**, *11*, 1667. [CrossRef]
11. Anguita, J.V.; Smith, C.T.G.; Stute, T.; Funke, M.; Delkowski, M.; Silva, S.R.P. Dimensionally and environmentally ultra-stable polymer composites reinforced with carbon fibres. *Nat. Mater.* **2019**, *19*, 317–322. [CrossRef] [PubMed]
12. Tiwari, S.; Bijwe, J. Surface Treatment of Carbon Fibers-A Review. *Procedia Technol.* **2014**, *14*, 505–512. [CrossRef]
13. Dai, Z.; Shi, F.; Zhang, B.; Li, M.; Zhang, Z. Effect of sizing on carbon fiber surface properties and fibers/epoxy interfacial adhesion. *Appl. Surf. Sci.* **2011**, *257*, 6980–6985. [CrossRef]
14. Zhang, G.; Sun, S.; Yang, D.; Dodelet, J.P.; Sacher, E. The surface analytical characterization of carbon fibers functionalized by H2SO4/HNO3 treatment. *Carbon* **2008**, *46*, 196–205. [CrossRef]
15. Choi, M.H.; Jeon, B.H.; Chung, I.J. The effect of coupling agent on electrical and mechanical properties of carbon fiber/phenolic resin composites. *Polymer* **2000**, *41*, 3243–3252. [CrossRef]
16. Pittman Jr, C.U.; Wu, Z.; Jiang, W.; He, G.-R.; Wu, B.; Li, W.; Gardner, S.D. Reactivities of amine functions grafted to carbon fiber surfaces by tetraethylenepentamine. Designing interfacial bonding. *Carbon* **1997**, *35*, 929–943. [CrossRef]
17. Becker-Staines, A.; Bremser, W.; Tröster, T. Poly(dimethylsiloxane) as Interphase in Carbon Fiber-Reinforced Epoxy Resin: Topographical Analysis and Single-Fiber Pull-Out Tests. *Ind. Eng. Chem. Res.* **2019**, *58*, 23143–23153. [CrossRef]
18. Cheng, X.; He, Z.; Luo, Y.; Zhang, X.; Lei, C. Manipulating Interfacial Strength of Polyphosphazene Functionalized Carbon Fiber Composites. *Polym. Compos.* **2019**, *40*, E1831–E1839. [CrossRef]
19. Zhang, X.; Xu, H.; Fan, X. Grafting of amine-capped cross-linked polyphosphazenes onto carbon fiber surfaces: A novel coupling agent for fiber reinforced composites. *Rsc Adv.* **2014**, *4*, 12198–12205. [CrossRef]
20. Liu, L.; Yan, F.; Li, M.; Zhang, M.; Xiao, L.; Ao, Y. Self-assembly of graphene aerogel on carbon fiber for improvement of interfacial properties with epoxy resin. *Mater. Lett.* **2018**, *218*, 44–46. [CrossRef]
21. Gao, B.; Zhang, R.; He, M.; Sun, L.; Wang, C.; Liu, L.; Zhao, L.; Cui, H.; Cao, A. Effect of a multiscale reinforcement by carbon fiber surface treatment with graphene oxide/carbon nanotubes on the mechanical properties of reinforced carbon/carbon composites. *Compos. Part A Appl. Sci. Manuf.* **2016**, *90*, 433–440. [CrossRef]
22. Keyte, J.; Pancholi, K.; Njuguna, J. Recent Developments in Graphene Oxide/Epoxy Carbon Fiber-Reinforced Composites. *Front. Mater.* **2019**, *6*, 224. [CrossRef]
23. Qin, Y.; Qu, M.; Pan, Y.; Zhang, C.; Schubert, D.W. Fabrication, characterization and modelling of triple hierarchic PET/CB/TPU composite fibres for strain sensing. *Compos. Part A Appl. Sci. Manuf.* **2020**, *129*, 105724. [CrossRef]
24. Zhao, M.; Meng, L.; Ma, L.; Ma, L.; Yang, X.; Huang, Y.; Ryu, J.E.; Shankar, A.; Li, T.; Yan, C.; et al. Layer-by-layer grafting CNTs onto carbon fibers surface for enhancing the interfacial properties of epoxy resin composites. *Compos. Sci. Technol.* **2018**, *154*, 28–36. [CrossRef]
25. Peng, Q.; Li, Y.; He, X.; Lv, H.; Hu, P.; Shang, Y.; Wang, C.; Wang, R.; Sritharan, T.; Du, S. Interfacial enhancement of carbon fiber composites by poly(amido amine) functionalization. *Compos. Sci. Technol.* **2013**, *74*, 37–42. [CrossRef]
26. Xu, H.; Zhang, X.; Liu, D.; Chun, Y.; Fan, X.; Chun. A high efficient method for introducing reactive amines onto carbonfiber surfaces using hexachlorocyclophosphazene as a new coupling agent. *Appl. Surf. Sci.* **2014**, *320*, 43–51. [CrossRef]
27. Islam, M.S.; Deng, Y.; Tong, L.; Faisal, S.N.; Roy, A.K.; Minett, A.I.; Gomes, V.G. Grafting carbon nanotubes directly onto carbon fibers for superior mechanical stability: Towards next generation aerospace composites and energy storage applications. *Carbon* **2016**, *96*, 701–710. [CrossRef]

28. Wang, L.; Liu, N.; Guo, Z.; Wu, D.; Chen, W.; Chang, Z.; Yuan, Q.; Hui, M.; Wang, J. Nitric Acid-Treated Carbon Fibers with Enhanced Hydrophilicity for Candida tropicalis Immobilization in Xylitol Fermentation. *Materials* **2016**, *9*, 206. [CrossRef]
29. Wu, T.; Wang, G.; Dong, Q.; Qian, B.; Meng, Y.; Qiu, J. Asymmetric capacitive deionization utilizing nitric acid treated activated carbon fiber as the cathode. *Electrochim. Acta* **2016**, *176*, 426–433. [CrossRef]
30. For comparison, TGA and Raman spectra of GO material and rGO aerogels have been recorded and analyzed in our previous work in ref [32].
31. Ganguly, A.; Sharma, S.; Papakonstantinou, P.; Hamilton, J. Probing the Thermal Deoxygenation of Graphene Oxide Using High-Resolution In Situ X-ray-Based Spectroscopies. *J. Phys. Chem. C* **2011**, *115*, 17009–17019. [CrossRef]
32. Vrettos, K.; Karouta, N.; Loginos, P.; Donthula, S.; Gournis, D.; Georgakilas, V. The Role of Diamines in the Formation of Graphene Aerogels. *Front. Mater.* **2018**, *5*. [CrossRef]
33. Staudenmaier, L. Verfahren zur Darstellung der Graphitsäure. *Ber. Der Dtsch. Chem. Ges.* **1898**, *31*, 1481–1487. [CrossRef]
34. Antonelou, A.; Sygellou, L.; Vrettos, K.; Georgakilas, V.; Yannopoulos, S.N. Efficient defect healing and ultralow sheet resistance of laser-assisted reduced graphene oxide at ambient conditions. *Carbon* **2018**, *139*, 492–499. [CrossRef]

Sample Availability: Samples of the compounds are available from the authors.

© 2020 by the authors. Licensee MDPI, Basel, Switzerland. This article is an open access article distributed under the terms and conditions of the Creative Commons Attribution (CC BY) license (http://creativecommons.org/licenses/by/4.0/).

Article

Thermal Conductivity of Defective Graphene Oxide: A Molecular Dynamic Study

Yi Yang [1], Jing Cao [2], Ning Wei [1,*], Donghui Meng [3], Lina Wang [3], Guohua Ren [3], Rongxin Yan [3,*] and Ning Zhang [1,*]

1. College of Water Resources and Architectural Engineering, Northwest A&F University, Yangling 712100, China; yang_yi@nwafu.edu.cn
2. State Key Laboratory of Eco-hydraulics in Northwest Arid Region of China, Xi'an University of Technology, Xi'an 710048, China; caojingxn@163.com
3. Beijing Institute of Spacecraft Environment Engineering, Beijing 100094, China; mengdonghui@126.com (D.M.); wangxiweigood@163.com (L.W.); wqghren@126.com (G.R.)
* Correspondence: nwei@nwsuaf.edu.cn (N.W.); pigsheepdog@126.com (R.Y.); johning@live.cn (N.Z.)

Academic Editors: Long Y Chiang and Giuseppe Cirillo
Received: 13 February 2019; Accepted: 12 March 2019; Published: 20 March 2019

Abstract: In this paper, the thermal properties of graphene oxide (GO) with vacancy defects were studied using a non-equilibrium molecular dynamics method. The results showed that the thermal conductivity of GO increases with the model length. A linear relationship of the inverse length and inverse thermal conductivity was observed. The thermal conductivity of GO decreased monotonically with an increase in the degree of oxidation. When the degree of oxidation was 10%, the thermal conductivity of GO decreased by ~90% and this was almost independent of chiral direction. The effect of vacancy defect on the thermal conductivity of GO was also considered. The size effect of thermal conductivity gradually decreases with increasing defect concentration. When the vacancy defect ratio was beyond 2%, the thermal conductivity did not show significant change with the degree of oxidation. The effect of vacancy defect on thermal conductivity is greater than that of oxide group concentration. Our results can provide effective guidance for the designed GO microstructures in thermal management and thermoelectric applications.

Keywords: graphene oxide; thermal conductivity; vacancy defect

1. Introduction

Graphene oxide (GO), an oxidation product of graphene [1], has attracted much attention in recent years as a two-dimensional material [2] because of its unique mechanical and thermal properties [3–5]. The structure of GO is composed of oxygen functional groups connected on the base plane of a layer of carbon atoms in two-dimensional space [1]. The existence of oxygen functional groups makes its thermal transport properties quite different from those of graphene. Graphene is the best known thermal conductive material. Its thermal conductivity can reach 2000–5000 W/mK [6]. However, the oxygen functional groups on the surface of GO destroy the lattice symmetry [7] and cause local strain [8], resulting in a reduction of thermal conductivity by 2–3 orders of magnitude [9]. Nika et al. indicated that the strong phonon scattering in GO resulted in a significant decrease in thermal conductivity [10].

On the other hand, the reduction method can further regulate the concentration of oxygen functional groups, which means the thermal transport properties of GO can be regulated in a larger range. Considering the size effect, Lin and Mu calculated the effect of different degrees of oxidation on the thermal conductivity of GO [11], and revealed that the thermal conductivity converges to 8.8 W/mK [12]. In recent experiments, the thermal conductivity of GO varies from 2 to 1000 W/mK using different oxygen reduction methods [13–15]. GO can be used in various thermal management electronic devices [16], such as electronic cooling [17], thermal diodes [18] and thermal logic circuits [19] due to the ability to adjust thermal conductivity. In addition, GO also shows good thermoelectric properties [4,20]. Therefore, considering the potential applications of GO in thermal management and thermoelectric energy conversion, it is necessary to study the thermal conductivity of GO.

In the process of preparation and reduction of GO, structural damage and vacancy defect are inevitable. GO is often regarded as a monolayer graphene with both oxygen functional groups and vacancy defects [1]. Renteria et al. revealed that the thermal conductivity of GO films is anisotropic [21]. In recent years, some progress has been made in the study of GO thermal conductivity. Zhao considered the effect of various defects on thermal conductivity of GO strips with fixed length [22]. The thermal conductivity of materials depends on phonons, and the phonon scattering is enhanced by GO defects, thus reducing the thermal conductivity [15,23]. On the other hand, with the presence of oxygen functional groups and doping defects, the thermal conductivity may be further reduced [9]. However, current studies cannot accurately describe the coupling effect of degree of oxidation and vacancy defects on thermal conductivity. Quantitative analysis of this problem is necessary.

In this study, the thermal conductivity of GO is calculated based on the non-equilibrium molecular dynamics method. Considering the coupling effect of oxygen group concentration and the ratio of vacancy defects, the variation of in-plane thermal conductivity of monolayer GO is studied, and the empirical formula for the ratio of vacancy defect, degree of oxidation and thermal conductivity of GO is established.

2. Model and Methodology

GO has two main surface groups, hydroxyl and epoxy groups [24]. The main factor affecting the thermal conductivity of GO is the content of functional groups (degree of oxidation) rather than the type of functional groups [22]. Therefore, only one functional group type of hydroxyl (-OH) is considered in this work.

Here, GO with randomly distributed vacancy defects and hydroxyl groups was built as shown in Figure 1. To make the calculation model more consistent with the actual situation, the quenching process of GO was simulated using the ReaxFF reactive force field under NPT ensemble [25,26]. The GO, established with several different initial functional group concentrations, was first gradually heated from 300 to 500 K over a time span of 0.2 ns, then annealed at 500 K for 0.2 ns, and subsequently quenched to 300 K over a time span of 0.2 ns. Finally, the model was further annealed at 300 K and zero pressure for the duration of 0.2 ns to ensure complete equilibration of the structure. Thus, each GO model was obtained with the final functional group concentration after quenching.

Through the above steps, the hydroxyl groups and vacancy at several different ratios were introduced in the model. The hydroxyl groups were randomly attached to the carbon atoms on both sides of the graphene basal plane at different degrees of oxidation ranging from 0% to 10%, while removing the carbon atoms from the GO sheet on the surface defect from 0% to 2%.

In the present study, the dynamic response of the system shown in Figure 2 was revealed by a molecular dynamics (MD) approach. The MD simulations were carried out by using the large-scale atomic/molecular massively parallel simulator (LAMMPS) [27]. The all-atom optimized potential for liquid simulations (OPLS-AA) was used for the study of GO thermal conductivity to improve the computation efficiency [28–30].

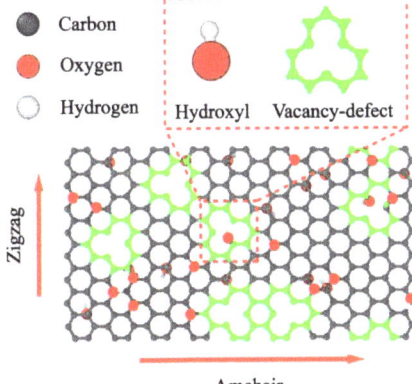

Figure 1. Schematic picture of graphene oxide (GO) with randomly distributed vacancy defects and hydroxyl groups.

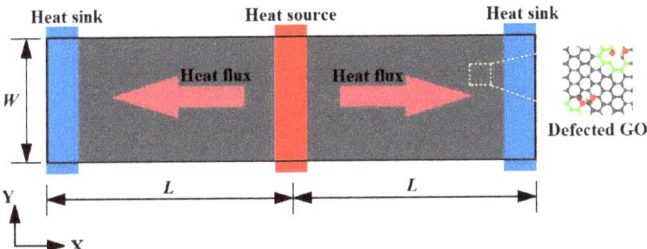

Figure 2. Schematic model for thermal conductance of GO using periodic boundary conditions.

To avoid the computational problems created by high frequency vibration caused by bond stretching energy (-OH) and bond angle bending energy (C-O-H), the SHAKE algorithm was adopted to fix atoms. Coulomb interactions were computed by using the particle-particle particle-mesh (PPPM) algorithm [31]. In this work, the thermal conductivity was computed by reverse non-equilibrium molecular dynamics (RNEMD) simulations in a microcanonical NVE ensemble [32]. The key point of the method is to impose a heat flux through the system and to determine the temperature gradient and temperature junctions as a consequence of the imposed flux. The fastest descent method was used to redistribute the atomic positions.

The above systems were equally divided into 100 thin slabs along the heat transfer direction, with the heat source and sink each taking one of the slabs. The heat source (hot slab) and sink (cold slab) slabs were located at the middle and the two ends of the model, respectively. The periodic boundary conditions were applied in the X and Y direction. A time step of 0.1 fs was selected for integration of the equations of atomic motion in the simulations. The system reached the equilibrium state at 300 K in Nosé-Hoover thermal bath for 0.2 ns. Then, the system was switch linear fitted to the NVE ensemble to exchange the kinetic energies (every 1000-time steps) between the coldest atom in the heat sink slab and the hottest atom in the heat source slab for 0.8 ns. The total heat flux J can be obtained from the amount of the injected/released two slabs by exchanging the kinetic energies Equation (1).

$$J = \frac{\sum_{N_{tranfers}} \frac{1}{2}(mv_h^2 - mv_c^2)}{t_{transfer}}, \quad (1)$$

where $N_{tranfers}$ is the total number of exchanging the kinetic energies, $t_{tranfers}$ is the time over which the exchanging simulation is started, m represents the mass of the atoms, v_h and v_c are the velocities of the hottest atom of the cold slab and the coldest atom of the hot slab, respectively. When the heat flow in the structure reaches the non-equilibrium steady state, the temperature profiles is collected to obtain the temperature gradient as Equation (2).

$$T_i = \frac{2}{3Nk_B}\sum_j \frac{p_j^2}{2m_j},\qquad(2)$$

where T_i is the temperature of the N number atoms in i-th slab. m_j, v_j and p_j represent the mass and velocity and momentum of the atom j in i-th slab, respectively. The term k_B is Boltzmann's constant. The temperature profiles are obtained by averaging results of the last 8 million timesteps.

Four typical samples of the temperature profiles of monolayer GO are shown in Figure 3, where the temperature gradient ∇T (dT/dx) was obtained by linear fitting in the linear region of the profile along the longitudinal heat flux direction in Figure 3. The thermal conductivity κ_G can be calculated as Equation (3).

$$\kappa_G = \frac{J}{2A\nabla T},\qquad(3)$$

where A is the cross-section area of corresponding models and the constant 2 in the denominator accounts for the fact that the system is periodic.

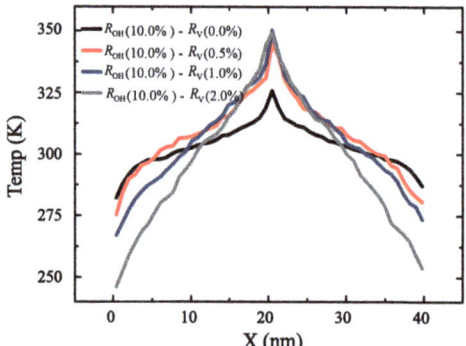

Figure 3. Schematic plot for reverse non-equilibrium molecular dynamics (RNEMD) simulations and equilibrium temperature profiles for GO.

3. Result and Discussion

First, the effects of the sample width on the thermal conductivity was investigated through MD simulations. As shown in Figure 4, in the range of 2 to 10 nm for different chirality with a fixed length of 20 nm, the measure of increasing width W acquired a convergent thermal conductivity.

Then, the effects of the sample length on the thermal conductivity (κ_G) along the zigzag and armchair directions were explored with the length varying from 20 to 180 nm and a fixed width of 2 nm. The results (see Figure 5) clearly show that the thermal conductivity does not depend on the sample's width. A linear relationship of the inverse length and inverse thermal conductivity can be observed. This means that the thermal conductivity increases with the length, two fitting functions are $\kappa_{G(Zigzag)}^{-1} = 0.4704L^{-1} + 0.00857$ and $\kappa_{G(Armchair)}^{-1} = 0.4697L^{-1} + 0.00856$.

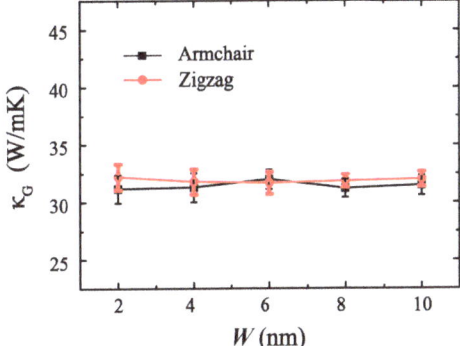

Figure 4. Curve of thermal conductivity with different sample width. The width varies in [0, 10] nm.

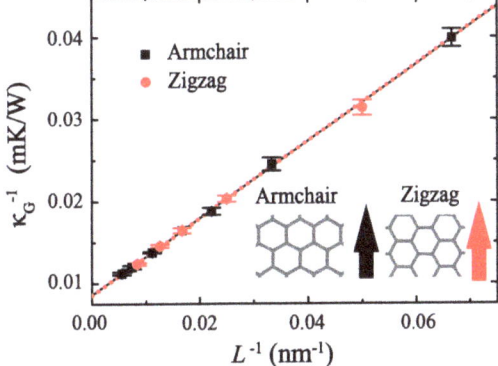

Figure 5. The relationship of length and thermal conductivity in GO (R_{OH}: ~10%) along zigzag (red) and armchair (black) directions at 300 K.

The relationship between κ_G^{-1} and L^{-1} can also be expressed as [33]:

$$\kappa_G^{-1} = \kappa_\infty^{-1}(\frac{2l}{L} + 1), \qquad (4)$$

where l is the mean free path (MFP) of phonon. κ_∞ denotes the thermal conductivity in infinite length. Through Equation (4), the thermal conductivity κ_∞ along the zigzag and armchair directions was found to be 116.82 and 116.68 W/mK, respectively. The corresponding MFP of phonon values l were 27.45 nm (along zigzag direction) and 27.44 nm (along armchair direction), which are much smaller than that of graphene (~775 nm) [6].

Through the classical lattice heat transport theory, the thermal conductivity of low-dimensional material can be calculated by $\kappa = Cvl$, where C is the specific heat, v is the group velocity. Previous literature has indicated that the values of C and v changed little by analyzing phonon density of states in GO [15]. This explains why the thermal conductivity of GO is smaller than that of graphene.

To study the coupling effects of the hydroxyl-group and vacancy defects on the thermal conductivity of GO, we defined a ratio between oxygen and carbon atoms R_{OH} to describe the degree of oxidation. Also, R_V is defined as the ratio of vacancy defect in the system, which can be calculated by the density of atoms removed from the pristine GO.

From Figure 6, the concentration of functional groups and the ratio of vacancy defects have a negative impact on the thermal conductivity of the structure in a certain degree. For a known concentration of functional groups, the thermal conductivity of the structure decreases gradually with the increase in vacancy defects in the structure. The decline in thermal conductivity is no longer obvious with the increase in vacancy defect ratio. When the vacancy defect $R_V \leq 1.0\%$, the thermal conductivity is very sensitive to both the change in vacancy defect and the concentration of functional groups. For GO without vacancy defect, the thermal conductivity drops most significantly while the functional group concentration increases. When R_V exceeds 2.0%, the functional group concentration has little effect on thermal conductivity.

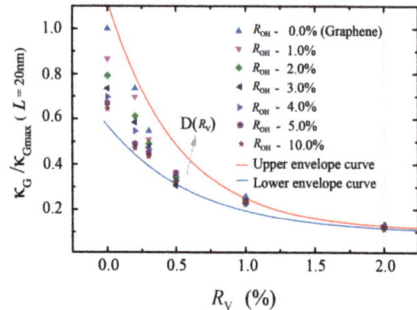

Figure 6. The relative thermal conductivity of GO with varying degrees of oxidation and vacancy defect ratio in the same sample size of 20 nm. Six different symbols indicate the different degree of oxidation with varied vacancy defect ratios, the red and blue line denote the fitting curves.

According to the results, the lower and upper envelope curves of nonlinear fitting are drawn in Figure 6. The upper curve in red indicates the thermal conductivity of the model is only affected by the vacancy defect ratio. The fitting formula is $\kappa_G/\kappa_{Gmax} = 0.1142 + 0.8859 e^{-R_V/0.5126}$. The lower envelope curve in blue is the thermal conductivity of the system with 10% oxidation affected by the vacancy defect ratio. The fitting formula is $\kappa_G/\kappa_{Gmax} = 0.1003 + 0.5034 e^{-R_V/0.6365}$. The region between the lower and upper envelope curve indicates all the cases of coupling effects between a single vacancy (R_V: 0 ~ 2%) and the hydroxyl group (R_{OH}: 0 ~ 10%) in 20 nm length (Figure 6). The simulation results also reveal that the effect of vacancy defects on thermal conductivity of GO is greater than that of functional group concentration.

To explore the coupling effect of such factors, we define the $D(R_V)$ (see Figure 6) as the difference between the upper envelope curve and lower envelope curve at a same ratio of vacancy. $D(R_V)$ decreases as R_V increases and approximately approaches zero when $R_V > 2.0\%$. Results indicated that the vacancy has a strong effect on thermal conductivity compared with the oxygen functional concentration. For example, when $R_V = 2.0\%$, the thermal conductivity with samples size of 20 nm is about 6.01 W/mK, regardless of the changing concentration of the functional group.

To further investigate the thermal conductivity on a macroscopic scale, the coupling effect of R_{OH} and R_V with five different GO lengths was employed. The ranges of the GO envelope are shown in black curves in Figure 7. As the length (L) of the GO sheet increases, the area between the lower and upper envelope curves is extended.

Combined with Equation (4), the thermal conductivity is extrapolated to infinite size. As the red curves show in Figure 7, the upper envelope indicates that the thermal conductivity of graphene tends to converge with the increase in the defect ratio. The results are similar to those obtained by Malekpour [34,35]. Also, the lower envelope is a thermal conductivity of R_{OH} = 10% GO. Two lines indicate that the maximum range of thermal conductivity can be up to 96%. With the increase of R_V, the regulatory range of functional groups decreases gradually. The range of functional group regulation is only ~11% when the vacancy defect ratio is at 1%. When the vacancy defect reaches 2%, the concentration of functional groups has little effect on the thermal conductivity. Therefore, in order to obtain a larger range of thermal conductivity control capabilities, it is necessary to reduce the vacancy defects in GO.

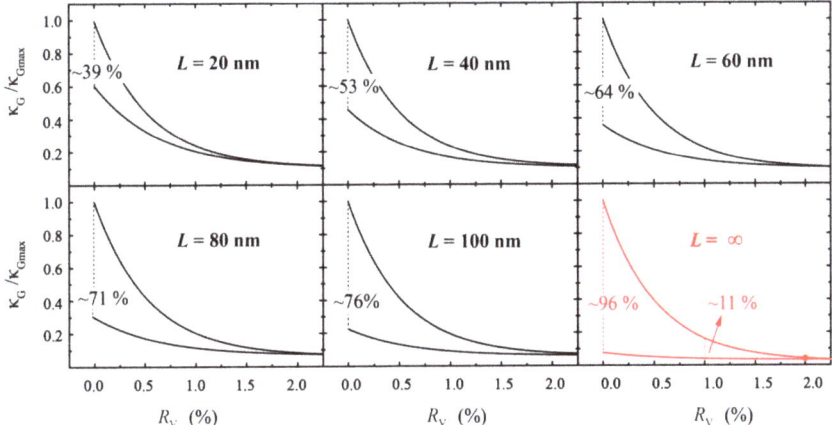

Figure 7. Relative thermal conductivity in different sample sizes.

As shown in Figure 8, with the increase in vacancy defects in GO, the size effect is no longer obvious. The thermal conductivity converges to 6.23 W/mK with a 2% defect. This proves that the thermal conductivity of defect-GO is less dependent on model length than that of the corresponding graphene and GO, since the thermal conductivity of defect-GO is mainly influenced by short-range acoustic and optical phonons which are length-independent [36]. Also, the less length-dependent thermal conductivity of defect-GO indicates that the long-range acoustic phonons are mainly scattered at vacancy. Moreover, a linear relationship of the inverse length and inverse thermal conductivity can be observed in the four types of defect ratio (see Figure 8b). Through formula (4), the corresponding MFPs of phonon are shown in Table 1. When the simulated size is larger than the MFP of phonon, the ballistic transport no longer plays a leading role and the thermal conductivity gradually converges [10]. Therefore, the larger the defect ratio, the smaller the simulation domain size as the GO thermal conductivity converges.

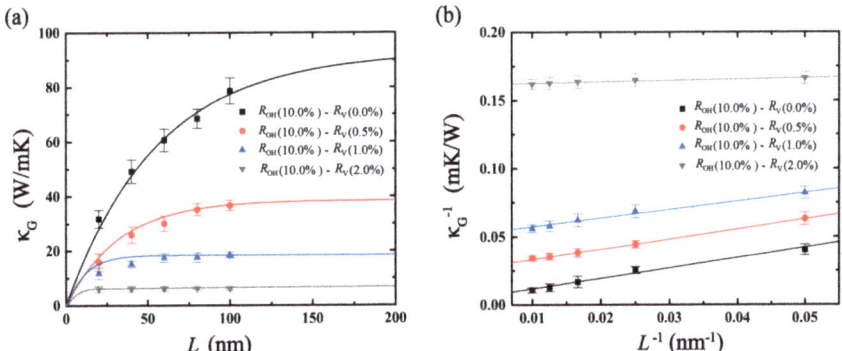

Figure 8. Length dependence of defect-GO's thermal conductivity. Solid lines are best fit to Equation (4). (**a**) The relationship between κ and L, (**b**) the relationship between κ^{-1} and L^{-1}.

To elucidate the mechanism of heat transfer of GO sheets, the spatial distribution of the heat flux by vector arrows on each atom under non-equilibrium steady state is shown in Figure 9, which displays the heat flux of GO for the specified structure.

The atomic heat flux is defined from the expression: $\vec{J_i} = e_i\vec{v_i} - S_i\vec{v_i}$, where e_i, v_i, and S_i are the energy, velocity vector and stress tensor of each atom i, respectively [37]. It can be obtained by calculating the atomic heat flux in the MD simulations and the results are averaged over 1 ns. The vector arrows show the migration of the heat flux on the GO and vividly reflect the transformation of the heat flux path as well as the phonon scattering around the vacancy/hydroxyl group regions.

Table 1. The mean free path (MFP) of phonon for four types of defect ratio in GO.

Type	Fitting Functions	MFP of Phonon
$R_{OH}(10\%) - R_V(0.0\%)$	$\kappa_G^{-1} = 0.4697L^{-1} + 0.00856$	27.44 nm
$R_{OH}(10\%) - R_V(0.5\%)$	$\kappa_G^{-1} = 0.4422L^{-1} + 0.02581$	8.57 nm
$R_{OH}(10\%) - R_V(1.0\%)$	$\kappa_G^{-1} = 0.2199L^{-1} + 0.05130$	2.14 nm
$R_{OH}(10\%) - R_V(2.0\%)$	$\kappa_G^{-1} = 0.0895L^{-1} + 0.15162$	0.29 nm

The heat flow scattering occurs at the vacancy and hydroxyl group regions on the surface of GO (see Figure 9). When a propagating heat flux tries to pass through a barrier in GO, under a single vacancy defect, the heat flow not only diffuses out of the plane, but also disturbs the heat flow around the pore in the plane. The heat flow shows irregular transmission while the addition of functional groups only slightly disturbs the surrounding heat flow. In other words, the hydroxyl groups do not break the underlying hexagonal lattice and preserve relatively well the lattice symmetry of carbon atoms and integrity, thus disturb the thermal transport weakly. In contrast, the presence of vacancies reduces the thermal conductivity of graphene significantly as they break the in-plane network of sp^2 carbon bonds. Therefore, among the factors affecting thermal conductivity, the scattering effect of functional groups is less than that of vacancy defects. As shown in the previous analysis, when the vacancy defect ratio reaches a certain value, the perturbation caused by functional groups is covered by vacancy defects and the influence is negligible, thus, the change in thermal conductivity with the concentration of functional groups is no longer obvious.

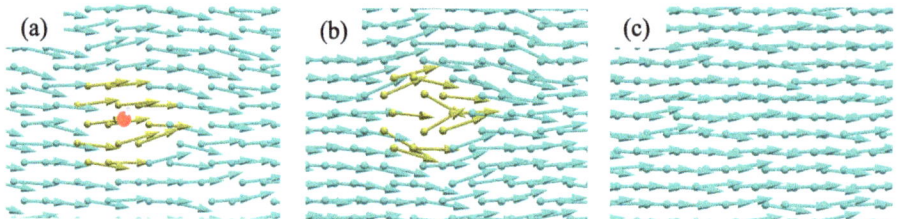

Figure 9. Spatial distribution of heat flux by vector arrows on each atom under non-equilibrium steady state. (a) A hydroxyl group, (b) one single vacancy, (c) graphene.

4. Conclusions

In summary, classical MD simulations were performed to investigate the thermal conductivity of GO with vacancy defect. Based on the simulation results, we found that GO has a significant size effect. The size effect of GO deteriorates with the increase in vacancy defects. It was also found that the effect of vacancy defects on thermal conductivity is more obvious than the degree of oxidation. With the increase in vacancy defects, the ability of functional group concentration to regulate the thermal conductivity of GO decreases. When the vacancy defect ratio is over 2%, the thermal conductivity does not show significant change with the degree of oxidation. This study provides theoretical guidance for the design and manufacture of thermoelectric and thermal management devices using GO as a raw material.

Author Contributions: Y.Y., D.M. and L.W. performed the simulations; N.W. and N.Z. conceived the simulations; G.R. and R.Y. conducted segmental data processing and J.C. and Y.Y. wrote the manuscript. All authors discussed and approved the final version

Funding: The research was financially support by the National Natural Science Foundation of China (Grants. 11502217 and U1537109), Natural Science Foundation of ShaanXi (Grant No. 2018JM1008 and No17JK0574), China Postdoctoral Science Foundation (No. 2015M570854 and 2016T90949), HPC of NWAFU, the Youth Training Project of Northwest A&F University (No. Z109021600) and the Fundamental Research Funds for the Central Universities (No. Z109021712).

Conflicts of Interest: The authors declare no competing financial interest.

References

1. Yanwu, Z.; Shanthi, M.; Weiwei, C.; Xuesong, L.; Won, S.J.; Potts, J.R.; Ruoff, R.S. Graphene and graphene oxide: Synthesis, properties, and applications. *Cheminform* **2010**, *22*, 3906–3924.
2. Stankovich, S.; Dikin, D.A.; Dommett, G.H.; Kohlhaas, K.M.; Zimney, E.J.; Stach, E.A.; Piner, R.D.; Nguyen, S.T.; Ruoff, R.S. Graphene-based composite materials. *Nature* **1990**, *442*, 282. [CrossRef] [PubMed]
3. Goki, E.; Giovanni, F.; Manish, C. Large-area ultrathin films of reduced graphene oxide as a transparent and flexible electronic material. *Nat. Nanotechnol.* **2008**, *3*, 270–274.
4. Choi, J.; Tu, N.D.K.; Lee, S.S.; Lee, H.; Jin, S.K.; Kim, H. Controlled oxidation level of reduced graphene oxides and its effect on thermoelectric properties. *Macromol. Res.* **2014**, *22*, 1104–1108. [CrossRef]
5. Tian, L.; Pickel, A.D.; Yao, Y.; Chen, Y.; Zeng, Y.; Lacey, S.D.; Li, Y.; Wang, Y.; Dai, J.; Wang, Y. Thermoelectric properties and performance of flexible reduced graphene oxide films up to 3000 K. *Nat. Energy* **2018**, *3*, 148–156.
6. Balandin, A.A.; Ghosh, S.; Bao, W.; Calizo, I.; Teweldebrhan, D.; Miao, F.; Lau, C.N. Superior thermal conductivity of single-layer graphene. *Nano Lett.* **2008**, *8*, 902. [CrossRef]
7. Zhang, C.; Dabbs, D.M.; Liu, L.M.; Aksay, I.A.; Car, R.; Selloni, A. Combined effects of functional groups, lattice defects, and edges in the infrared spectra of graphene oxide. *J. Phys. Chem. C* **2015**, *119*, 150720175209001. [CrossRef]
8. Nekahi, A.; Marashi, S.P.H.; Fatmesari, D.H. Modified structure of graphene oxide by investigation of structure evolution. *Bull. Mater. Sci.* **2015**, *38*, 1717–1722. [CrossRef]
9. Baek, S.J.; Hong, W.G.; Min, P.; Kaiser, A.B.; Kim, H.J.; Kim, B.H.; Park, Y.W. The effect of oxygen functional groups on the electrical transport behavior of a single piece multi-layered graphene oxide. *Synth. Met.* **2014**, *191*, 1–5. [CrossRef]
10. Nika, D.L.; Balandin, A.A. Phonons and thermal transport in graphene and graphene-based materials. *Rep. Prog. Phys.* **2016**, *80*, 036502. [CrossRef]
11. Lin, S.; Buehler, M.J. Thermal transport in monolayer graphene oxide: Atomistic insights into phonon engineering through surface chemistry. *Carbon* **2014**, *77*, 351–359. [CrossRef]
12. Mu, X.; Wu, X.; Zhang, T.; Go, D.B.; Luo, T. Thermal transport in graphene oxide-from ballistic extreme to amorphous limit. *Sci. Rep.* **2014**, *4*, 3909. [CrossRef]
13. Timo, S.; Burg, B.R.; Schirmer, N.C.; Dimos, P. An electrical method for the measurement of the thermal and electrical conductivity of reduced graphene oxide nanostructures. *Nanotechnology* **2009**, *20*, 405704.
14. Mahanta, N.K.; Abramson, A.R. Thermal conductivity of graphene and graphene oxide nanoplatelets. *Therm. Thermomechanic. Phenom. Electron. Syst.* **2012**. [CrossRef]
15. Zhang, H.; Fonseca, A.F.; Cho, K. Tailoring thermal transport property of graphene through oxygen functionalization. *J. Phys. Chem. C* **2015**, *118*, 1436–1442. [CrossRef]
16. Kargar, F.; Barani, Z.; Lewis, J.S.; Debnath, B.; Balandin, A.A. Thermal percolation threshold and thermal properties of composites with graphene and boron nitride fillers. *ACS Appl. Mater. Interfaces* **2018**, *10*, 37555–37565. [CrossRef] [PubMed]
17. Chang, C.W.; Okawa, D.; Majumdar, A.; Zettl, A. Solid-state thermal rectifier. *Science* **2006**, *314*, 1121–1124. [CrossRef]
18. Baowen, L.; Lei, W.; Giulio, C. Thermal diode: Rectification of heat flux. *Phys. Rev. Lett.* **2004**, *93*, 184301.
19. Hu, J.; Ruan, X.; Chen, Y.P. Thermal conductivity and thermal rectification in graphene nanoribbons: A molecular dynamics study. *Nano Lett.* **2009**, *9*, 2730. [CrossRef]
20. Wang, W.; Zhang, Q.; Li, J.; Liu, X.; Wang, L.; Zhu, J.; Luo, W.; Jiang, W. An efficient thermoelectric material: Preparation of reduced graphene oxide/polyaniline hybrid composites by cryogenic grinding. *Rsc Adv.* **2015**, *5*, 8988–8995. [CrossRef]

21. Renteria, J.D.; Ramirez, S.; Malekpour, H.; Alonso, B.; Centeno, A.; Zurutuza, A.; Cocemasov, A.I.; Nika, D.L.; Balandin, A.A. Anisotropy of thermal conductivity of free-standing reduced graphene oxide films annealed at high temperature. *Adv. Funct. Mater.* **2015**, *25*, 4664–4672. [CrossRef]
22. Zhao, W.; Wang, Y.; Wu, Z.; Wang, W.; Bi, K.; Liang, Z.; Yang, J.; Chen, Y.; Xu, Z.; Ni, Z. Defect-engineered heat transport in graphene: A route to high efficient thermal rectification. *Sci. Rep.* **2015**, *5*, 11962. [CrossRef] [PubMed]
23. Sheng, C.; Zhang, Y.; Huang, Q.; Hao, W.; Wang, G. Effects of vacancy defects on graphene nanoribbon field effect transistor. *Micro Nano Lett.* **2013**, *8*, 816–821.
24. He, H.; Klinowski, J.; Forster, M.; Lerf, A. A new structural model for graphite oxide. *Chem. Phys. Lett.* **1998**, *287*, 53–56. [CrossRef]
25. Kimberly, C.; Duin, A.C.T.; van Goddard, W.A. ReaxFF reactive force field for molecular dynamics simulations of hydrocarbon oxidation. *J. Phys. Chem. A* **2008**, *112*, 1040–1053.
26. Medhekar, N.V.; Ashwin, R.; Ruoff, R.S.; Shenoy, V.B. Hydrogen bond networks in graphene oxide composite paper: Structure and mechanical properties. *ACS Nano* **2010**, *4*, 2300–2306. [CrossRef] [PubMed]
27. Plimpton, S. Fast parallel algorithms for short-range molecular dynamics. *J. Comput. Phys.* **1995**, *117*, 1–19. [CrossRef]
28. Jorgensen, W.L.; Maxwell, D.S.; Tirado-Rives, J. Development and testing of the OPLS All-Atom force field on conformational energetics and properties of organic liquids. *J. Am. Chem. Soc.* **1996**, *118*, 11225–11236. [CrossRef]
29. Wei, N.; Lv, C.; Xu, Z. Wetting of graphene oxide: A molecular dynamics study. *Langmuir Acs J. Surf. Colloids* **2014**, *30*, 3572. [CrossRef] [PubMed]
30. Cao, J.; Cai, K. Thermal expansion producing easier formation of a black phosphorus nanotube from nanoribbon on carbon nanotube. *Nanotechnology* **2017**, *29*. [CrossRef]
31. Hockney, R.W.; Eastwood, J.W. *Computer Simulation Using Particles*; Taylor & Francis, Inc.: Milton Park, UK, 1988.
32. Müllerplathe, F. A simple nonequilibrium molecular dynamics method for calculating the thermal conductivity. *J. Chem. Phys.* **1997**, *106*, 6082–6085. [CrossRef]
33. Zhang, Y.Y.; Pei, Q.X.; Jiang, J.W.; Wei, N.; Zhang, Y.W. Thermal conductivities of single-and multi-layer phosphorene: A molecular dynamics study. *Nanoscale* **2015**, *8*, 483–491. [CrossRef]
34. Malekpour, H.; Ramnani, P.; Srinivasan, S.; Balasubramanian, G.; Nika, D.L.; Mulchandani, A.; Lake, R.K.; Balandin, A.A. Thermal conductivity of graphene with defects induced by electron beam irradiation. *Nanoscale* **2016**, *8*, 14608–14616. [CrossRef]
35. Malekpour, H.; Balandin, A.A.; Malekpour, H.; Balandin, A.A. Raman-based technique for measuring thermal conductivity of graphene and related materials: Thermal conductivity of graphene and related materials. *J. Raman Spectrosc.* **2018**, *49*, 106–120. [CrossRef]
36. Yang, L.; Chen, J.; Yang, N.; Baowen, L. Significant reduction of graphene thermal conductivity by phononic crystal structure. *Int. J. Heat Mass Transf.* **2015**, *91*, 428–432. [CrossRef]
37. Wei, N.; Chen, Y.; Cai, K.; Zhao, J.; Wang, H.-Q.; Zheng, J.-C. Thermal conductivity of graphene kirigami: Ultralow and strain robustness. *Carbon* **2016**, *104*, 203–213. [CrossRef]

Sample Availability: Samples of the compounds are available from the authors.

© 2019 by the authors. Licensee MDPI, Basel, Switzerland. This article is an open access article distributed under the terms and conditions of the Creative Commons Attribution (CC BY) license (http://creativecommons.org/licenses/by/4.0/).

Systematic Investigations of Annealing and Functionalization of Carbon Nanotube Yarns

Maik Scholz [1,2], Yasuhiko Hayashi [3,*], Victoria Eckert [1,2], Vyacheslav Khavrus [1,†], Albrecht Leonhardt [1], Bernd Büchner [1,4], Michael Mertig [2,5] and Silke Hampel [1,*]

1. Leibniz Institute for Solid State and Material Research Dresden, Helmholtzstr. 20, 01069 Dresden, Germany; maik_scholz@web.de (M.S.); victoria-eckert@web.de (V.E.); vhavrus@gmail.com (V.K.); leoalb@web.de (A.L.); B.Buechner@ifw-dresden.de (B.B.)
2. Institute for Physical Chemistry, Technische Universität Dresden, 01062 Dresden, Germany; michael.mertig@tu-dresden.de
3. Graduate School of Natural Science and Technology, Okayama University, 3-1-1 Tsushima-naka, Kita, Okayama 700-8530, Japan
4. Institute for Solid State Physics, Technische Universität Dresden, 01062 Dresden, Germany
5. Kurt-Schwabe-Institut für Mess- und Sensortechnik e.V. Meinsberg, 04736 Waldheim, Germany
* Correspondence: hayashi.yasuhiko@okayama-u.ac.jp (Y.H.); S.Hampel@ifw-dresden.de (S.H.); Tel.: +81-(0)86-251-8230 (Y.H.); +49-(0)351-4659-323 (S.H.)
† Current address: Life Science Inkubator Sachsen GmbH & Co. KG, Tatzberg 47, 01307 Dresden, Germany.

Academic Editors: Long Y Chiang and Giuseppe Cirillo
Received: 11 December 2019; Accepted: 28 February 2020; Published: 4 March 2020

Abstract: Carbon nanotube yarns (CNY) are a novel carbonaceous material and have received a great deal of interest since the beginning of the 21st century. CNY are of particular interest due to their useful heat conducting, electrical conducting, and mechanical properties. The electrical conductivity of carbon nanotube yarns can also be influenced by functionalization and annealing. A systematical study of this post synthetic treatment will assist in understanding what factors influences the conductivity of these materials. In this investigation, it is shown that the electrical conductivity can be increased by a factor of 2 and 5.5 through functionalization with acids and high temperature annealing respectively. The scale of the enhancement is dependent on the reducing of intertube space in case of functionalization. For annealing, not only is the highly graphitic structure of the carbon nanotubes (CNT) important, but it is also shown to influence the residual amorphous carbon in the structure. The promising results of this study can help to utilize CNY as a replacement for common materials in the field of electrical wiring.

Keywords: carbon nanotube yarns; carbon nanotube; functionalization; electrical conductivity; annealing; acid treatment

1. Introduction

To fulfill future claims on everyday applications for higher efficiency, new materials with improved physical properties and lower production costs are necessary. One of the most promising candidates for these materials is carbon nanotubes (CNT). These one-dimensional tubular carbon structures have been shown to possess an impressive array of physical properties on the scale of individual tubes [1], including but not limited to ballistic electron transport [2], high thermal conductivity [3] and mechanical strength [4]. Together with their low density and high current carrying capacity [5], they outperform most commonly used materials such as copper. Translating these outstanding properties from the nano to the bulk scale is therefore understandably a major focus in the field of CNT research. Besides CNT hybrid systems with polymers, spinning macroscopic yarns out of different CNT starting materials like sheets and arrays is a promising way to incorporate these properties into practical materials. These carbon nanotube yarns (CNY) have a high tensile strength [6], are extremely flexible and very light weight, which makes these materials promising candidates for electrical wiring in different applications [7–11]. However the physical properties, especially the electrical conductivity, still lag behind individual CNT and copper [1].

The main factors defining the electrical conductivity of CNY are the intrinsic electrical properties of the CNT and the contact resistant between adjacent CNT. Different groups have shown that adjusting the CNT type used for spinning can improve the electrical properties of these yarns. An ideal CNY would consist purely out of long metallic single walled CNT and reach conductivity as high as an individual CNT [12]. Other ways to enhance the electrical properties of CNY include post-spinning treatments such as annealing and chemical functionalization. Annealing is known to repair structural defects in CNT [13,14] and is a reagent free way to improve the electrical properties of CNY [15,16]. Different approaches to anneal CNY have been developed over the years [15,17,18]. However, because CNT are only connected through weak van der Waals forces where functionalization by an acid treatment can only influence the inter-tube interactions by means of surface modification. It has been claimed that introducing oxygen rich groups raise the charge carrier density between adjacent CNT [19,20]. There is an ongoing controversy if functional groups are the main factor for increasing electrical conductivity or the change of yarn diameter through this acid treatment [21,22].

In this work, we provide a systematical study on the effects of annealing and functionalization by acid treatment to CNT yarns. Hereby different acids were tested for functionalization as well as annealing temperatures up to 2500 °C.

2. Results and Discussion

2.1. Annealing

The pure multiwalled CNT (MWCNT) array was shown to consist of CNT with 2 to 6 walls with the majority of CNT possessing 2 to 4 walls [23]. The diameter of pristine yarn was around 22 ± 1 µm (Figure 1 and Figure S1).

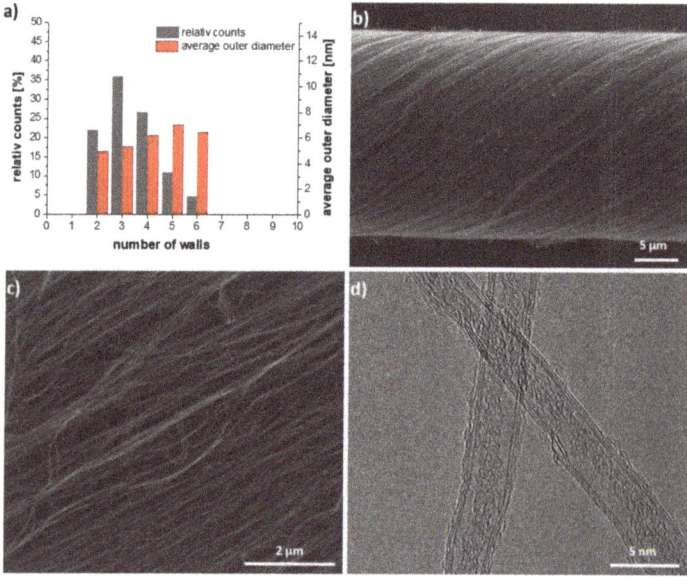

Figure 1. (**a**) Statistical analysis of the pristine carbon nanotube (CNT), (**b**,**c**) SEM images of pristine carbon nanotube yarn (CNY) and (**d**) TEM image of pristine CNT.

For improvement of the properties of CNY and therefore the electrical conductivity the CNY were annealed under an Argon atmosphere at 1000, 1500, 2000 and 2500 °C for 30, 60 and 120 min (Figure 2). The appearance of all CNY after annealing differs from the untreated yarn with a higher density surface, lower spacing and fewer voids between the CNT bundles (Figure 2c). Additionally, the diameter decreased about 1 µm. After annealing at 2500 °C CNT with 4 to 6 walls make up the majority—over 70 % of the observable CNT (Figure 2a and Figure S2). The increase in the number of walls is a process that has also been observed in other works [24]. This behavior can be explained by the increased thermal activity of the outer wall and the desire of the system to take a state of energetic minimum. When examined with TEM, annealed CNT show a much straighter structure compared to that of the pristine yarn (Figure 2d). There are also fewer recognizable defects in the wall structure. This is consistent with reports that high-temperature treatment mainly influences the microstructure of the CNT or CNT walls [25].

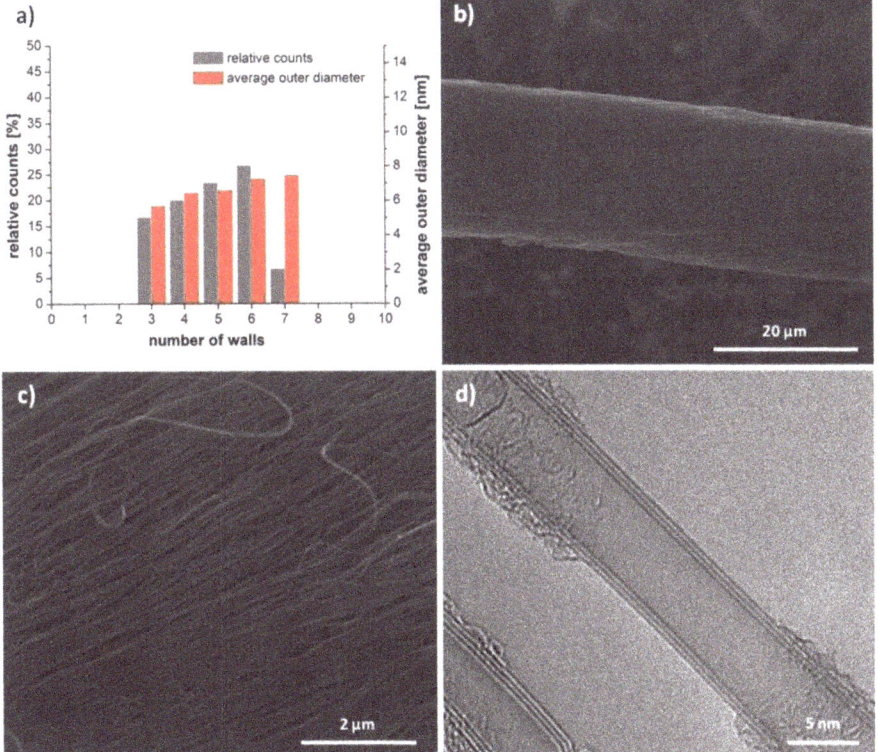

Figure 2. Investigation of CNY annealed for 2 h at 2500 °C. (**a**) Statistical analysis of the annealed CNT, (**b**,**c**) SEM images of annealed CNY and (**d**) TEM image of annealed CNT.

The improvement in CNY microstructure is also supported by Raman measurements. With increasing time and temperature the I_D/I_G ratio drops from 0.83 to 0.23 for samples annealed at 2500 °C for 2 h (Figures 3 and 4).

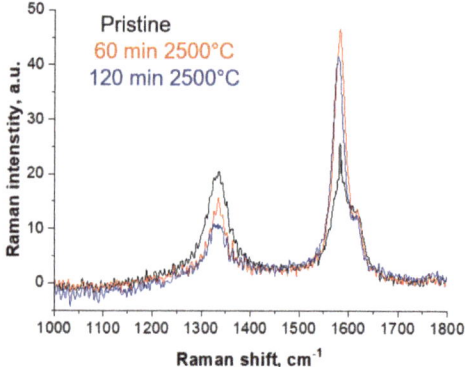

Figure 3. Raman spectra of pristine CNY and of annealed CNY for 60 and 120 min at 2500 °C.

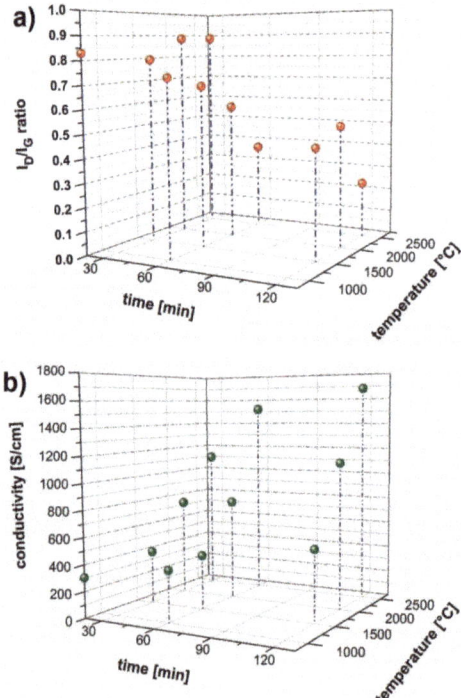

Figure 4. (a) Raman spectroscopy I_D/I_G ratio as a function of annealing time and temperature, (b) electrical conductivity of CNT yarns as function of annealing time and temperature.

Increased graphitization becomes noticeable after 1 h annealing at 1500 °C with a decrease of the I_D/I_G ratio to 0.68. After 2 h of annealing at 1500 °C the ratio drops to 0.45, which corresponds to almost 50% of the untreated yarn sample. This shows that it is not only temperature, but also the time of annealing that influences the increasing the graphitization of this material (Figures 3 and 4).

Improved crystallinity via high temperature annealing is also in good agreement with previously reported results for CNT yarns [15,17] and individual CNT [13,14]. Comparing the results of this study to previous work [17], it is suggested that other annealing techniques such as resistance-heating in vacuum might improve the crystallinity even more. This may be linked to the fact that the conventional annealing used for this work does not result in the observable evaporation of amorphous carbon from the sample. This fact may play an important role in the quality of the treated samples, as we will discuss below in the results of electrical conductivity.

Looking at the behavior of the electrical conductivity, it is noticeable that in the same ratio as the I_D/I_G ratio is decreasing the conductivity at room temperature increases (Figure 4a,b). Electrical conductivity is enhanced from 300 S/cm for pristine yarn to ~1680 S/cm for the yarn sample annealed for 2 h at 2500 °C. This is a significant increase of ~460% in total for a single treatment.

Because of the higher crystallinity of the sample, it is thought that a lack of defects in the CNT walls and therefore less scattering points for electrons within the individual CNT might be related to this increase in conductivity. But another aspect is here also relevant. Comparing these results to our previous work, we still find amorphous carbon within the CNT after annealing. This carbon will function as a conductive bridging agent between adjacent highly graphitic CNT (Figure 5) and provide a pathway for conduction of electrons. According to the so-called Mott variable range hopping (VRH) model, the electrons jump from one starting point to another with the lowest possible hopping energy. The electrons must overcome the gap between one CNT and the other. Amorphous carbon is therefore helping to bridge this process. The whole process a CNY is an interplay between a high crystalline and high conductive CNT and the next CNT. Here a compressed CNY structure (like after annealing or acid functionalisation) as well as additional carbon as a conductive bridge is ideal to overcome the gaps by hopping [26].

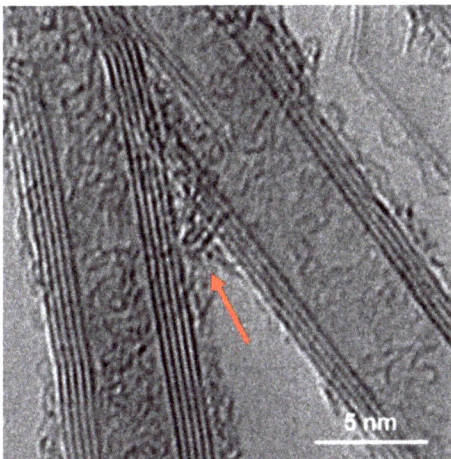

Figure 5. Bridging of adjacent CNT by amorphous carbon after annealing at 2500 °C for 2 h.

Figure 6a,b show specific and normalized electrical conductivity of CNT yarns annealed for 2 h at different temperatures in the range of 5 to 295 K. Each sample shows monotonically increasing conductivity with temperature, which is typical for MWCNT materials [27–29]. The electrical conductivity also increases over the whole temperature range with increasing annealing temperature. In comparison to Kaiser et al. [30] and Skákalová et al. [31] the curves of normalized conductivity indicate that the predominant conduction mechanism for both the pristine and annealed CNY is three-dimensional variable range hopping. This mechanism describes a phonon-assisted tunneling process between localized charge carrier states [32,33]. It is, therefore, thought that the hopping takes place between occupied and unoccupied states that are separated both spatially and energetically from each other. As the temperature of the sample decreases, the thermal energy of the phonons recede and fewer states become energetically attainable, thus the electrical conductivity drops. With increasing annealing temperature the normalized temperature dependent conductivity measurements show a slight decrease of the slope with increasing annealing temperature (Figure 6b).

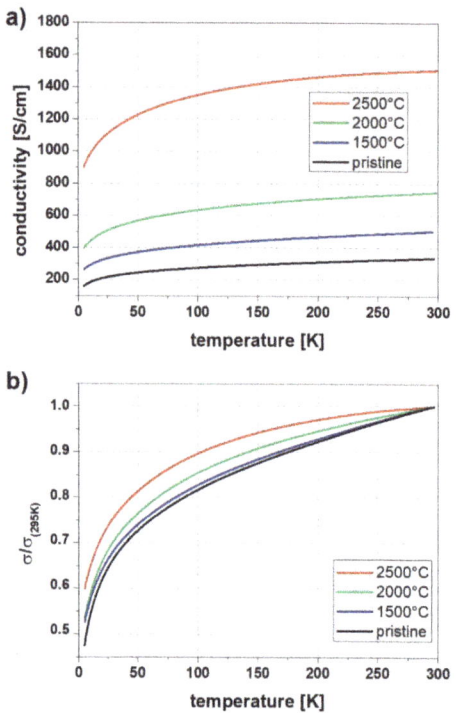

Figure 6. (a) Specific and (b) normalized electrical conductivity of pristine and 2 h annealed CNT yarns.

2.2. Acid Treatment

To further increase the electrical conductivity the CNY were functionalized. A variety of different acids were used and the acid treatment was carried out at room temperature for 3 to 216 h for each of the acid treatment processes.

Oxidative acids like HNO_3 have been used to purify CNT and introduce functional groups on their surface. Diluted nitric acid has been used to purify CNT [34] whereas concentrated or mixtures of strong acids effectively introduce oxygen rich groups onto CNT [35]. The effects of acidification time on the performance of CNT yarns was examined, with different high concentrated acids, namely H_2O_2 (30%), HCl (37%), HNO_3 (65%) and half concentrated HNO_3 (32%).

After acid treatment, the appearance of the yarn differs strongly depending on the kind of acid and treatment time. The least difference in appearance between the pristine yarn and the acid treated were observed after the H_2O_2 treatment. Even after 216 h, the yarn structure does not differ significantly from the original sample (Figure S3). Due to capillary forces during drying of the yarns after treatment no loose CNT bundles are noticeable. After 216 h HCl treatment leads to a deformation of the yarn in the direction of twisting. Short treatment times, however, do not lead to any significant change in the appearance (Figure S4). In the case of treatment with 50% concentrated HNO_3, the CNY shows clear deformations and indentations along the direction of twist after a short treatment time of 3 h. Increasing treatment time furthers the deformation of the yarn (Figure 7). Nevertheless, the surface remains closed and without voids or gaps (Figure 7b,d). As with H_2O_2 and HCl, the capillary forces occurring during drying after acid treatment are the cause. It should also be noted that the twist of the yarn is maintained under these conditions even after a treatment time of 216 h.

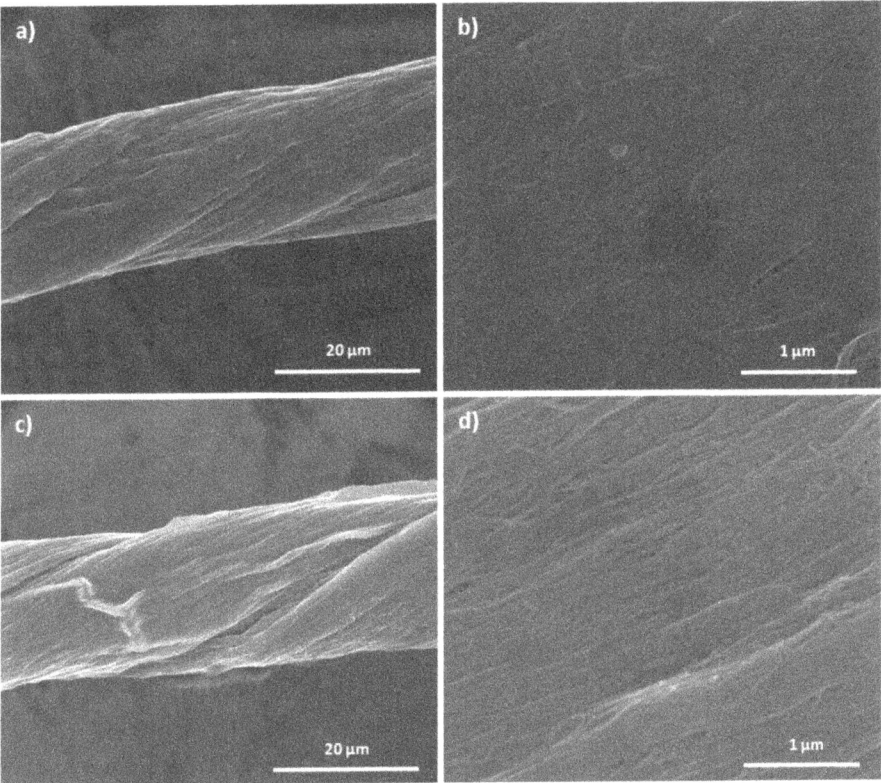

Figure 7. SEM images of yarns after treatment with half concentrated HNO_3: (**a,b**) after 3 h and (**c,d**) after 216 h treatment time.

Treatment with concentrated HNO_3 results in a dramatic change of appearance. Even after a short treatment time, significant indentations and deformations of the yarn are observable. The twist of the yarn is already no longer recognizable after 3 h treatment. However, the surface of the yarn remains dense and closed as with the treatments by the other acids. No major gaps or similar defects appear. As the treatment time increases, the deformation of the yarn increases and in addition to the depressions along the fiber direction, the yarn also vertically folds perpendicular to the fiber direction occur (Figure 8). However, after a treatment time of 216 h some, but not all, of the outer layer of the yarn appears to have broken off (Figure 8c inset). This indicated that the outer layer is no longer as tightly bound at the remaining inner part of the yarn but only a quarter of the examined length shows this behavior. The other sections of the yarn surface appear to remain closed, as with treatment with the other acids (Figure 8d).

These results show that CNT yarns are attacked and deformed differently by different acid treatments. There is a ranking of H_2O_2, HCl, half and concentrated HNO_3, where concentrated HNO_3 causes the largest morphology changes. One cause of these deformations is thought to be insufficient compaction of the yarn during spinning. This may leave voids inside the yarn structure, which could be are compressed by the surface tension of the acids or contracted by the capillary forces during the drying of the yarns.

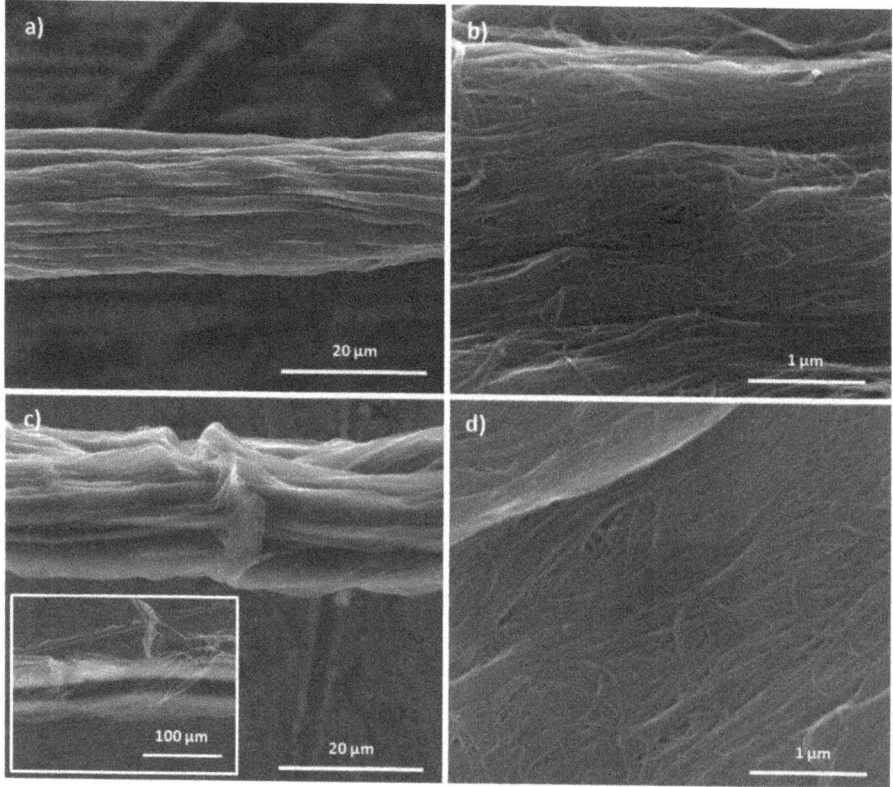

Figure 8. SEM images of yarns after treatment with concentrated HNO_3: (**a,b**) after 3 h and (**c,d**) after 216 h treatment time.

In addition to the effect of acid treatment on the surface and the appearance of the yarn, the effects on the structure, diameter and electrical conductivity of the yarn were investigated. Figure 9 shows the results of these investigations for the different acid treatments as a function of time. One striking feature common to all the acids treatments is that the maximum increase of conductivity is achieved with a treatment time of 3 h. With increasing treatment time the conductivity drops to lower values and in the case of concentrated HNO_3 even to the value of the untreated yarn (300 S/cm).

The behavior of the yarn after treatment with concentrated HNO_3 is different from the other tested acids. There is a strong variation of the I_D/I_G ratio but only a relatively small fluctuation of the average yarn diameter observed with this treatment. However, it should be noted that the measured diameter of the yarn is subject to high fluctuations, as indicated by the large error bars. These variations are due to the severe deformation and partial detachment of parts of the outer yarn layer. Over the considered experimental period (3 to 216 h), the I_D/I_G ratio is subject to large variations ranging from 0.68 to 0.88. This, in turn, shows the strong impact of concentrated HNO_3 on the structure of CNT materials. Electrical conductivity drops after the strong increase at 3 h and drops to 300 S/cm after 216 h. The maximum value for the conductivity is 642 S/cm, which represents an increase of more than 110%. For verification, a trial was carried out with only one hour of treatment, which only resulted in a value of 416 S/cm. With this, it appears that 3 h represents an optimal treatment time with concentrated HNO_3 for the yarn used in this work. The drop of conductivity after 3 h treatment could be explained by the strong oxidative nature of concentrated HNO_3, this results in a strong defect introducing behavior. This appears to be proof that the action of introducing functional groups on the surface of CNY is not as effective as claimed as increasing treatment time should result in a noticeable positive effect on conductivity by increasing the charge carrier density between the CNT. In addition to this, a stable densification of the yarn through the functional groups and introduced stronger dipol-dipol and H-bridge bonds should be observable [20].

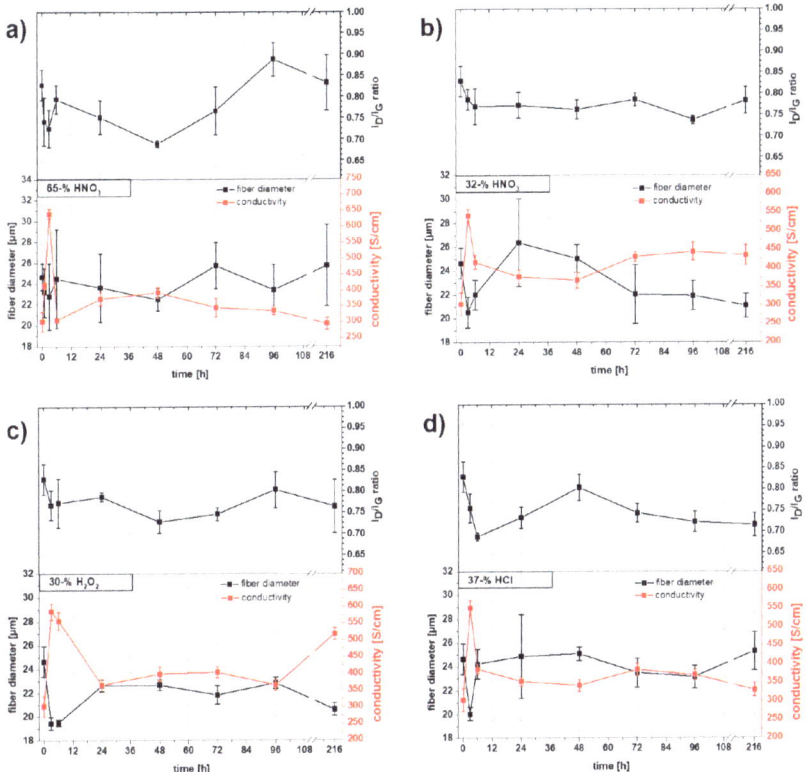

Figure 9. Conductivity, I_D/I_G ratio and fiber diameter of CNY after different acidification times: (**a**) concentrated nitric acid, (**b**) half concentrated nitric acid, (**c**) concentrated hydrogen peroxide and (**d**) concentrated hydrochloric acid.

The influence of the other acids (H_2O_2, HCl and half concentrated HNO_3) is less dramatic than with concentrated HNO_3, with the I_D/I_G ratio being only minimally affected. It is even decreased by the influence of HCl. After 216 h, the values are in the range of 0.72 to 0.78. This decrease is explained in literature by the removal of carbonaceous impurities during acid treatment [36,37]. However, the nature of these carbonaceous impurities was not explained. It has been observed in this work that the yarn diameter and the electrical conductivity appears to show a strong correlation. This is best seen in the half concentrated HNO_3 treated sample with the electrical conductivity increases with decreasing diameter and vice versa. The maximum values reached after 3 h are for H_2O_2, HCl and half concentrated HNO_3 587, 551 and 541 S/cm, respectively. In the case of H_2O_2 a rise in electrical conductivity (518 S/cm) can be observed after a treatment time of 216 h. However this effect has not yet been explained and requires further investigation. It might be assumed that temperature fluctuations and the influence of light during the experiment leads to a decomposition of H_2O_2 and formation of hydroxyl radicals that can promote the formation of functional groups on the surface of the CNT.

Different groups describe a similar positive effect on the electrical conductivity of CNT yarns by acid treatment [20–22,36,38]. However, the explanation for the increase in conductivity is different in each case. One of the most common explanations is that the influence of oxidative acids such as nitric acid or mixtures of nitric acid and other acids functionalizes the surface of the CNT with oxygen-containing groups such as hydroxyl, carboxyl and carbonyl groups. This increases the electron density between the CNT and creates additional conduction paths between the CNT [20]. Another explanation is that the smaller distance between the CNT after acid treatment reduces the contact resistance, by which the electron-hopping mechanism is supported [22]. Our experiments support the hypothesis of Meng et al. [22] where the formation of functional groups plays a secondary role compared to the compaction of the yarns. However, Meng et al. have described this effect only for concentrated HNO_3. In our experiments, we show that other acids lead to a similar effect. Because of the hydrophobic nature of the CNT yarns, water-based solutions like the used acids can only minimally infiltrate the inner structure of the yarns when treated for a short time. It is therefore thought that the yarns get compressed from the outside by the surface tension of the solutions. This results in the inter-tube spacing becoming smaller, thereby reducing the contact resistance between CNT. A less significant effect was observed with water. Here the shrinkage of diameter and rise of conductivity isn't as high as with the acids. This could be caused by the slight difference in surface tension for high concentrated acids [38–40] in comparison to water (ca. 73 mN/m at 20 °C). After the initial compression, the yarn diameter again increases nearly to the value of pristine yarn. It is thought that the acids slowly infiltrate the inner yarn structure with treatment time and widen the spaces between CNT bundles. However, as these bundles are difficult to infiltrate, they remain compressed even over long time, which would explain the remaining higher conductivity for half-concentrated HNO_3, H_2O_2 and HCl compared to pristine yarn.

The temperature-dependent conductivity measurements after acid treatment (Figure 10) support the theory of yarn compression during acid treatment. There is no change of the slope of the curves after 3 h and 216 h for half-conc. HNO_3, H_2O_2 and HCl. Only concentrated HNO_3 shows a slight influence on the conduction mechanism, presumably due to its oxidative nature. After 216 h an increase of the slope indicates the decomposition of the CNT.

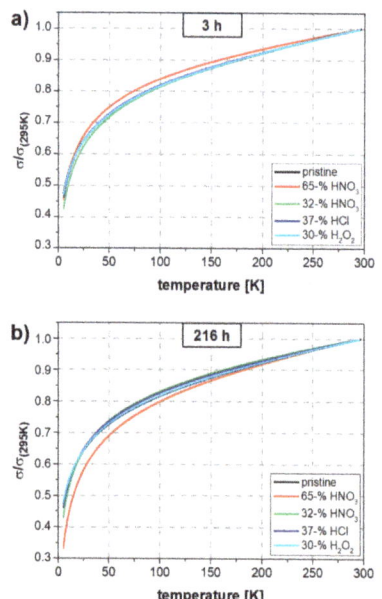

Figure 10. Normalized electrical conductivity of pristine yarn and yarns after 3 (**a**) and 216 h (**b**) of acid treatment.

3. Materials and Methods

CNY were produced by a two-step dry spinning process from a multiwalled CNT (MWCNT) array. The MWCNT array were grown by chemical vapor deposition as described by Iijima et al. [23]. The parameters for spinning the yarn were 1000 turns per min with a spinning speed of 40 mm/min. The resulting yarns were determined by thermogravimetric analysis TGA and energy-dispersive X-ray spectroscopy (EDX) measurements to be free of impurities including catalyst particles. We could not verify any further elements.

Annealing of CNY was conducted in a high temperature furnace (LHTG 100-200/30-1G, GERO, Neuhausen, Germany) under an Argon atmosphere and normal pressure. For this treatment, CNY were put into closed graphite vessels. Sections of the yarns were annealed at 1000, 1500, 2000 and 2500 °C for 30, 60 and 120 min.

For functionalization, a variety of different acids were used including concentrated (65-% or 14.4 mol/L; VWR AnalaR Normapur, Dresden, Germany) and half concentrated (32-% or 6.05 mol/L; VWR AnalaR Normapur, Dresden, Germany) nitric acid, conc. hydrochloric acid (37-% or 12.02 mol/L; VWR AnalaR Normapur, Dresden, Germany) and conc. hydrogen peroxide (30-% or 9.8 mol/L; Sigma Aldrich, Darmstadt, Germany). Pieces of 3 cm long CNY were put into glass vials with an excess amount of acid (5–10 mL). The acid treatment was carried out at 25 °C for 3 to 216 h for each of the acid treatment processes. Directly after each treatment, the samples were washed with water and dried over night at 108 °C.

CNY were characterized by SEM (Nova NanoSEM 200, FEI) and TEM (TITAN, FEI) before and after each annealing and acid treatment. SEM and TEM investigations were conducted using a cathode voltage of 15 kV and 80 kV respectively. For TEM investigations we spread the CNY in a mechanical way using tweezers to individualize several CNT. These CNY samples were fixed on a copper grid with special TEM glue. We investigated serval parts of the samples on the basis of more than 20 SEM images and about 75 TEM images. For statistical analysis of the CNT diameter a well as the number of walls we evaluated were min. 100 CNT per sample (Figures S1 and S2).

Characterization with a Micro-Raman Spectrometer (Horiba Jobin Yvon, France) was performed over between 1000 and 1800 cm^{-1} using a wavelength of 514.5 nm (Argon-Laser, Coherent, Santa Clara, CA 95054, USA). I_D/I_G ratios were calculated from Raman spectra by dividing the intensity of the D-band through the intensity of the G-band. For each sample, 5 Raman measurements were conducted. Electrical conductivity was measured between 5 and 295 K with a Nanovoltmeter (Keithley Instruments, Solon OH44139, USA) using the four-point measurement method. Here the current will be operated by two contacts at the ends of the CNY whereas the voltage will be measured by two additional contacts, which are located between the two current supply contacts (Figure S5). Specific electrical conductivity (σ) was calculated by the equation $\sigma = \frac{l}{RA}$ where l is the length between the inner contacts of the four-point set up, A is the cross-sectional area of the CNY calculated from the diameter measured from SEM images and R is the measured resistance.

4. Conclusions

In this work, we show a systematic study of the influence of annealing and functionalization by treatment with highly concentrated acids on the electrical conductivity and structure of CNY. Annealing enhances electrical conductivity by a factor of more than 5.5. A high graphitization of the CNT at 2500 °C leads to enhanced transport of electrons through the individual CNT. It was found the amorphous carbon resulting from the synthesis of the CNT plays an important role by connecting the CNT in the yarn structure and helps to reduce the contact resistance between adjacent CNT. Acid treatment over longer times and with different kinds of acids leads to an increase of electrical conductivity with a treatment time of 3 h resulting in the optimal increase. With this method, an increase of more than two times can be achieved. Furthermore, it was found that this increase is less dependent on forming functional groups on the surface of the CNT, but on the compression of the yarns and reduction of the intertube space. These results for annealing and functionalization help to understand the influences on CNT yarns by different post synthetic treatments and reveal key factors that could assist in producing highly conductive CNT yarns.

Supplementary Materials: The following are available online at http://www.mdpi.com/1420-3049/25/5/1144/s1, Figure S1: TEM images of pristine CNY.; Figure S2: TEM images of annealed CNY for 2 h at 2500 °C; Figure S3: SEM images of CNY after treatment with H_2O_2: (a,b) after 3 h and (c,d) after 216 h treatment time, Figure S4: SEM images of CNY after treatment with HCl: (a,b) after 3 h and (c,d) after 216 h treatment time; Figure S5: detailed schematic illustration of the sample holder with fixed CNY for four-point measurements.

Author Contributions: Conceptualization, M.S., V.K., A.L. and S.H.; methodology, Y.H. and V.K.; validation, M.S. and S.H.; investigation, V.E.; resources, Y.H. and B.B.; data curation, A.L. and B.B.; writing—original draft preparation, M.S. and S.H.; writing—review and editing, Y.H., V.K. and M.M.; supervision, Y.H., S.H., M.M. and B.B. All authors have read and agreed to the published version of the manuscript.

Funding: This research was partially funded by JSPS KAKENHI Grant Numbers 18H01708 and 18H05208.

Acknowledgments: The authors want to thank Sandra Schiemenz for Raman investigations and Robert Heider for performing the annealing experiments.

Conflicts of Interest: The authors declare no conflicts of interest. The funders had no role in the design of the study; in the collection, analyses, or interpretation of data; in the writing of the manuscript, or in the decision to publish the results.

References

1. Lekawa-Raus, A.; Patmore, J.; Kurzepa, L.; Bulmer, J.; Koziol, K. Electrical properties of carbon nanotube based fibers and their future use in electrical wiring. *Adv. Funct. Mater.* **2014**, *24*, 3661–3682. [CrossRef]
2. Charlier, J.; Blase, X.; Roche, S. Electronic and transport properties of nanotubes. *Rev. Mod. Phys.* **2007**, *79*, 677–732. [CrossRef]
3. Kim, P.; Shi, L.; Majumdar, A.; McEuen, P.L. Thermal transport measurements of individual multiwalled nanotubes. *Phys. Rev. Lett.* **2001**, *87*, 215502. [CrossRef] [PubMed]
4. Gao, G.; Çagin, T.; Goddard, W.A. Energetics, structure, mechanical and vibrational properties of single-walled carbon nanotubes. *Nanotechnology* **1998**, *9*, 184–191. [CrossRef]
5. Hong, S.; Myung, S. A flexible approach to mobility. *Nat. Nanotechnol.* **2007**, *2*, 207–208. [CrossRef]
6. Zhang, X.; Li, Q.; Tu, Y.; Li, Y.; Coulter, J.Y.; Zheng, L.; Zhao, Y.; Jia, Q.; Peterson, D.E.; Zhu, Y.; et al. Strong carbon-nanotube fibers spun from long carbon-nanotube arrays. *Small* **2007**, *3*, 244–248. [CrossRef]
7. Davis, V.A.; Parra-Vasquez, A.N.G.; Green, M.J.; Rai, P.K.; Behabtu, N.; Prieto, V.; Booker, R.D.; Schmidt, J.; Kesselman, E.; Zhou, W.; et al. True solutions of single-walled carbon nanotubes for assembly into macroscopic materials. *Nat. Nanotechnol.* **2009**, *4*, 830–834. [CrossRef]
8. Behabtu, N.; Young, C.C.; Tsentalovich, D.E.; Kleinerman, O.; Wang, X.; Ma, A.W.K.; Bengio, E.A.; Waarbeek, R.F.T.; de Jong, J.J.; Hoogerwerf, R.E.; et al. Strong, Light, Multifunctional Fibers of Carbon Nanotubes with Ultrahigh Conductivity. *Science* **2013**, *339*, 182–186. [CrossRef]
9. Orbaek, A.W.; Aggarwal, N.; Barron, A.R. The development of a 'process map' for the growth of carbon nanomaterials from ferrocene by injection CVD. *J. Mater. Chem. A* **2013**, *1*, 14122–14132. [CrossRef]
10. Khanbolouki, P.; Tehrani, M. Viscoelastic behaviour of carbon nanotube yarns and twisted coils. In Proceedings of the ASME International Mechanical Engineering Congress and Exposition, Pittsburgh, PA, USA, 9–15 November 2018; Volume 12. V012T11A031.
11. Mclean, B.; Eveleens, C.A.; Mitchell, I.; Webber, G.B.; Page, A.J. Catalytic CVD synthesis of boron nitride and carbon nanomaterials—Synergies between experiment and theory. *Phys. Chem. Chem. Phys.* **2017**, *19*, 26466–26494. [CrossRef]
12. Xu, F.; Sadrzadeh, A.; Xu, Z.; Yakobson, B.I. Can carbon nanotube fibers achieve the ultimate conductivity? Coupled-mode analysis for electron transport through the carbon nanotube contact. *J. Appl. Phys.* **2013**, *114*, 063714. [CrossRef]
13. Huang, W.; Wang, Y.; Luo, G.; Wei, F. 99.9% purity multi-walled carbon nanotubes by vacuum high-temperature annealing. *Carbon* **2003**, *41*, 2585–2590. [CrossRef]
14. Yamamoto, G.; Shirasu, K.; Nozaka, Y.; Sato, Y.; Takagi, T.; Hashida, T. Structure–property relationships in thermally-annealed multi-walled carbon nanotubes. *Carbon* **2014**, *66*, 219–226. [CrossRef]

15. Niven, J.F.; Johnson, M.B.; Juckes, S.M.; White, M.A.; Alvarez, N.T.; Shanov, V. Influence of annealing on thermal and electrical properties of carbon nanotube yarns. *Carbon* **2016**, *99*, 485–490. [CrossRef]
16. Pierlot, A.; Woodhead, A.L.; Church, J.S. Thermal annealing effects on multi-walled carbon nanotube yarns probed by Raman spectroscopy. *Spectrochim. Acta Part A Mol. Biomol. Spectrosc.* **2014**, *117*, 598–603. [CrossRef] [PubMed]
17. Scholz, M.; Hayashi, Y.; Khavrus, V.; Chujo, D.; Inoue, H.; Hada, M.; Leonhardt, A.; Büchner, B.; Hampel, S. Resistance-heating of carbon nanotube yarns in different atmospheres. *Carbon* **2018**, *133*, 232–238. [CrossRef]
18. Gong, T.; Zhang, Y.; Liu, W.; Wei, J.; Wang, K.; Wu, D.; Zhong, M.; Zhong, M. Connection of macro-sized double-walled carbon nanotube strands by current-assisted laser irradiation. *J. Laser Appl.* **2008**, *20*, 122–126. [CrossRef]
19. Skakalova, V.; Kaiser, A.B.; Dettlaff-Weglikowska, U.; Hrnčariková, K.; Roth, S. Effect of chemical treatment on electrical conductivity, infrared absorption, and raman spectra of single-walled carbon nanotubes. *J. Phys. Chem. B* **2005**, *109*, 7174–7181. [CrossRef]
20. Morelos-Gomez, A.; Fujishige, M.; Vega-Díaz, S.M.; Ito, I.; Fukuyo, T.; Cruz-Silva, R.; Tristán-López, F.; Fujisawa, K.; Fujimori, T.; Futamura, R.; et al. High electrical conductivity of double-walled carbon nanotube fibers by hydrogen peroxide treatments. *J. Mater. Chem. A* **2016**, *4*, 74–82. [CrossRef]
21. Liu, P.; Hu, D.C.M.; Tran, T.Q.; Jewell, D. Electrical property enhancement of carbon nanotube fibers from post treatments. *Colloids Surf. A Physicochem. Eng. Asp.* **2016**, *509* (Suppl. C), 384–389. [CrossRef]
22. Meng, F.; Zhao, J.; Ye, Y.; Zhang, X.; Li, Q. Carbon nanotube fibers for electrochemical applications: Effect of enhanced interfaces by an acid treatment. *Nanoscale* **2012**, *4*, 7464–7468. [CrossRef]
23. Iijima, T.; Oshima, H.; Hayashi, Y.; Suryavanshi, U.B.; Hayashi, A.; Tanemura, M. In-situ observation of carbon nanotube fiber spinning from vertically aligned carbon nanotube forest. *Diam. Relat. Mater.* **2012**, *24*, 158–160. [CrossRef]
24. Endo, M.; Hayashi, T.; Muramatsu, H.; Kim, Y.; Terrones, H.; Terrones, M.; Dresselhaus, M.S. Coalescence of double-walled carbon nanotubes: Formation of novel carbon bicables. *Nano Lett.* **2004**, *4*, 1451–1454. [CrossRef]
25. Andrews, R.; Jacques, D.; Qian, D.; Dickey, E.C. Purification and structural annealing of multiwalled carbon nanotubes at graphitization temperatures. *Carbon* **2001**, *39*, 1681–1687. [CrossRef]
26. Shklovskii, B.I.; Efros, A.L. Variable-range hopping conduction. In *Electronic Properties of Doped Semiconductors*; Springer: Berlin/Heidelberg, Germany, 1984; pp. 202–227.
27. Pöhls, J.-H.; Johnson, M.B.; White, M.A.; Malik, R.; Ruff, B.; Jayasinghe, C.; Schulz, M.J.; Shanov, V. Physical properties of carbon nanotube sheets drawn from nanotube arrays. *Carbon* **2012**, *50*, 4175–4183. [CrossRef]
28. Jakubinek, M.; Johnson, M.B.; White, M.A.; Jayasinghe, C.; Li, G.; Cho, W.; Schulz, M.J.; Shanov, V. Thermal and electrical conductivity of array-spun multi-walled carbon nanotube yarns. *Carbon* **2012**, *50*, 244–248. [CrossRef]
29. Jakubinek, M.; White, M.A.; Li, G.; Jayasinghe, C.; Cho, W.; Schulz, M.J.; Schulz, M.J.; Shanov, V. Thermal and electrical conductivity of tall, vertically aligned carbon nanotube arrays. *Carbon* **2010**, *48*, 3947–3952. [CrossRef]
30. Kaiser, A.; Skákalová, V.; Roth, S. Modelling conduction in carbon nanotube networks with different thickness, chemical treatment and irradiation. *Phys. E Low Dimens. Syst. Nanostructures* **2008**, *40*, 2311–2318. [CrossRef]
31. Skákalová, V.; Kaiser, A.B.; Dettlaff-Weglikowska, U.; Hrnčariková, K.; Roth, S. Electronic transport in carbon nanotubes: From individual nanotubes to thin and thick networks. *Phys. Rev. B* **2006**, *74*, 085403. [CrossRef]
32. Khan, Z.; Husain, M.; Perng, T.P.; Salah, N.; Habi, S. Electrical transport via variable range hopping in an individual multi-wall carbon nanotube. *J. Phys. Condens. Matter* **2008**, *20*, 475207. [CrossRef]
33. Mott, N. Conduction in non-crystalline materials. *Philos. Mag. A J. Theor. Exp. Appl. Phys.* **1969**, *19*, 835–852. [CrossRef]
34. Hu, H.; Zhao, B.; Itkis, M.E.; Haddo, R.C. Nitric acid purification of single-walled carbon nanotubes. *J. Phys. Chem. B* **2003**, *107*, 13838–13842. [CrossRef]
35. Zhang, J.; Zou, H.; Qing, Q.; Yang, Y.; Li, Q.; Liu, Z.; Guo, X.; Du, Z. Effect of chemical oxidation on the structure of single-walled carbon nanotubes. *J. Phys. Chem. B* **2003**, *107*, 3712–3718. [CrossRef]
36. Janas, D.; Vilatela, A.C.; Koziol, K.K.K. Performance of carbon nanotube wires in extreme conditions. *Carbon* **2013**, *62*, 438–446. [CrossRef]

37. Liu, C.H.; Fan, S.S. Effects of chemical modifications on the thermal conductivity of carbon nanotube composites. *Appl. Phys. Lett.* **2005**, *86*, 123106. [CrossRef]
38. Martin, E.; George, C.; Mirabel, P. Densities and surface tensions of $H_2SO_4/HNO_3/H_2O$ solutions. *Geophys. Res. Lett.* **2000**, *27*, 197–200. [CrossRef]
39. Nasr-El-Din, H.A.; Al-Othman, A.M.; Taylor, K.C.; Al-Ghamdi, A.H. Surface tension of HCl-based stimulation fluids at high temperatures. *J. Pet. Sci. Eng.* **2004**, *43*, 57–73. [CrossRef]
40. Hawley, G. *Hawley's Condensed Chemical Dictionary*, 15th ed.; Wiley: New York, NY, USA, 2007.

Sample Availability: Samples of the compounds CNY and different modifications are available from the authors.

© 2020 by the authors. Licensee MDPI, Basel, Switzerland. This article is an open access article distributed under the terms and conditions of the Creative Commons Attribution (CC BY) license (http://creativecommons.org/licenses/by/4.0/).

Article

Synthesis and Intramolecular Energy- and Electron-Transfer of 3D-Conformeric Tris(fluorenyl-[60]fullerenylfluorene) Derivatives

He Yin [1], Min Wang [1], Loon-Seng Tan [2] and Long Y. Chiang [1,*]

1. Department of Chemistry, University of Massachusetts Lowell, Lowell, MA 01854, USA; He_Yin@student.uml.edu (H.Y.); wangmin81@gmail.com (M.W.)
2. Functional Materials Division, AFRL/RXA, Air Force Research Laboratory, Wright-Patterson Air Force Base, Dayton, OH 45433, USA; loon.tan@us.af.mil
* Correspondence: Long_Chiang@uml.edu; Tel.: +1-978-934-3663; Fax: +1-978-934-3013

Received: 3 August 2019; Accepted: 10 September 2019; Published: 13 September 2019

Abstract: New 3D conformers were synthesized to show a nanomolecular configuration with geometrically branched 2-diphenylaminofluorene (DPAF-C_{2M}) chromophores using a symmetrical 1,3,5-triaminobenzene ring as the center core for the connection of three fused DPAF-C_{2M} moieties. The design led to a class of *cis-cup*-tris[(DPAF-C_{2M})-C_{60}(>DPAF-C_9)] 3D conformers with three bisadduct-analogous <C_{60}> cages per nanomolecule facing at the same side of the geometrical molecular *cis-cup*-shape structure. A sequential synthetic route was described to afford this 3D configured conformer in a high yield with various spectroscopic characterizations. In principle, a nanostructure with a non-coplanar 3D configuration in design should minimize the direct contact or π-stacking of fluorene rings with each other during molecular packing to the formation of fullerosome array. It may also prevent the self-quenching effect of its photoexcited states in solids. Photophysical properties of this *cis-cup*-conformer were also investigated.

Keywords: Tri[60]fullerenyl stereoisomers; *cis-cup*-form of 3D-stereoisomers; tris(diphenylaminofluorene); 3D-configured nanostructures; intramolecular energy transfer for singlet oxygen production; intramolecular electron transfer for superoxide radical production

1. Introduction

Photoinduced intramolecular energy and electron transfer phenomena in organo [60]fullerene derivatives having a covalent molecular composition of both an electron donor and a [60]fullerenyl or nanocarbon acceptor components were demonstrated over a number of years [1–3]. The energy process may involve the facile triplet state of fullerene and other chromophores [4–8]. This type of nanomolecular system was used in many technological applications [9,10], including photovoltaic devices [11,12], sensors and switches [13], and photodynamic therapy [14,15]. Fullerene-based nanostructures with multiple C_{60} cages [16] in the structure were found to be suitable for the applications of nanocars [17–19], photoswitches [20], molecular heterojunctions [21], and catalysts [22]. Unusual molecular properties of multi-cage fullerene objects were theoretically predicted [23–26]. Recently, similar photophysical chemistry was also simulated in the modulation of photoswitchable dielectric properties to observe a large amplification of dielectric constants in a material combination form of multi-layered core-shell nanoparticles (NPs) [27–29]. The latter system was based on photoinduced intramolecular electronic charge-polarization of light-harvesting chromophoric nano[60]fullerenyl conjugates, such as 9,9-di(3,5,5-trimethylhexyl)-2-diphenylaminofluorenyl-methano[60]fullerene C_{60}(>DPAF-C_9) (**1**-C_9, Figure 1). The polarization provided detectable dielectric property enhancement in a layered [60]fullerosome membrane structure on gold-shelled nanoparticles. The phenomena were based on the high electronegativity of C_{60}> cage making it possible to rapidly shift an electron from light-harvesting DPAF (diphenylaminofluorene) donor moiety within the molecular structure to the C_{60}> moiety. This resulted in the formation of the corresponding charge-separated (CS) state $C_{60}^{-}\cdot(>DPAF-C_9)^+$ as the source of polarized charges. In fact, ultrafast concurrent intramolecular energy and electron transfer kinetics within **1**-C_9 were substantiated previously by femtosecond transient absorption measurements (pump-probe) [30–32]. When these negative charges were distributed, delocalized, and stabilized in the fullerosome membrane array at the shell layer on core-shell NPs, it resulted in a CS state with a lifetime prolonged enough for the detection of dielectric characteristics. The process involved interlayer photoinduced plasmonic energy transfer from the Au shell layer to the outer shell layer of $C_{60}(>DPAF-C_9)$ in addition to the fact that **1**-C_9 itself is also photoresponsive and excitable under light irradiation.

One crucial parameter to consider is the method of molecular packing within the fullerosome shell layer. In this regard, strong tendency of light-harvesting chromophores to aggregate among π-conjugated planar aromatic moieties can cause either concentration-dependent self-quenching effects of excited states or luminescence in the solid phase, including fullerosome. The π–π stacking may result in significant reduction of many photophysical properties. This type of packing aggregation can be partially minimized by the use of highly bulky and geometrically hindered π-conjugated chromophores in a structural design to restrict or distort intramolecular rotation bonding units with steric hindrance [33]. In the case of DPAF-C_9, we recently developed and synthesized highly restricted 3D conformers based on inter-connected three DPAF-C_9 chromophore units giving a structure of tris(DPAF-C_9) (**2**-C_9) [34] to prevent and minimize the tendency of planar DPAF units to undergo aromatic–aromatic stacking, overlapping, and aggregation via intermolecular hydrophobic–hydrophobic interactions in solid thin-films. This structural modification led the enhancement of photophysical properties, including the intensity of photoluminescence (PL) and electroluminescence (EL) emissions [35].

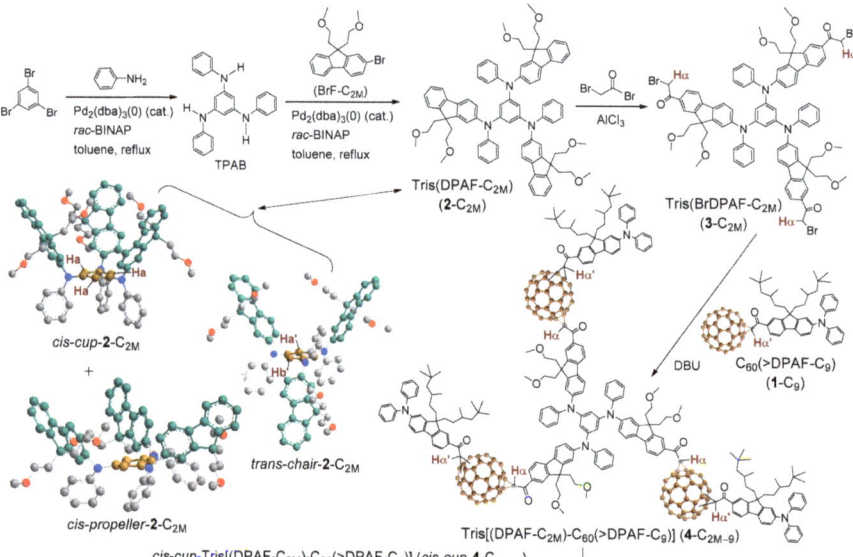

Figure 1. Synthetic methods for the preparation of 3D conformers of tris[(DPAF-C_{2M})-$C_{60}(>DPAF-C_9)$] (**4**-C_{2M-9}) with reaction reagents given and three perspective 3D-configurations of the key stereoisomeric intermediate **2**-C_{2M}.

Accordingly, we extended the similar structural strategy to design new 3D conformers tris[(DPAF-C$_{2M}$)-C$_{60}$(>DPAF-C$_9$)] (**4**-C$_{2M-9}$), as shown in Figure 1, for the study. The stereochemical modification was based on the construction of 3D geometrically branched tris[(DPAF-C$_{2M}$) (**2**-C$_{2M}$) chromophore having a shared central benzene unit among three 2-diphenylaminofluorenes. Highly steric hindrance at the corresponding 1,3,5-phenylfluorenylaminobenzene moiety forced three fluorene ring moieties to twist either upward or downward away from the central benzene plane with a large torsional angle on the nitrogen atom. This resulted in the formation of three to four possible stereoisomeric configurations, such as *cis-cup*-**2**-C$_{2M}$, *trans-chair*-**2**-C$_{2M}$, and *cis/or trans-propeller*-**2**-C$_{2M}$, as shown in Figure 1. All of these conformer forms were proposed to be capable of fully eliminating the tendency of **4**-C$_{2M-9}$ in inducing π–π aromatic–aromatic type stacking packing and to allow all C$_{60}$> cages to interact with each other via strong hydrophobic–hydrophobic interaction forces between (C$_{60}$>)–(C$_{60}$>) fullerenyl cages, forming the nano-layer array of fullerosome membrane.

2. Results and Discussion

Rapidly responsive nanophotonic physical properties of 3D conformers tris[(DPAF-C$_{2M}$)-C$_{60}$(>DPAF-C$_9$)] (**4**-C$_{2M-9}$) are achievable by specifically associating a donor–acceptor type chemical structure to a photomechanism having the ability to create a largely enhanced intramolecular energy and electron transfer efficiency. This mechanism occurs between C$_{60}$> acceptor and DPAF-C$_{2M}$/and DPAF-C$_9$ donor moieties bonded on fullerenes. In our functional group design of **4**-C$_{2M-9}$, a methanoketo bridging unit was used to trigger the keto-enol isomerization tautomerism that is capable of inducing π-periconjugation between C$_{60}$> and DFAP to provide a partial conjugation pathway for enhancing the π-electron mobility around the conjugated system of molecular nanostructures [36,37]. In addition, the new molecular design of stereoisomeric tris(fluorenylphenylamino)benzene [tris(DPAF-C$_9$)] analogous was proven to act as a fluorophore showing high intensity of photoluminescence and electroluminescence emission efficiency [34,35]. This revealed the successful utilization of stereochemistry to allow hindered and branched 3′,5′,5′-trimethylhexyl (C$_9$) arms to maintain the space-separation of three planar DPAF moieties intramolecularly within the nanostructure. It also behaved similarly in intermolecular packing that improved light-harvesting efficiency.

Based on the molecular formation energy based density functional theory (DFT) calculations of three plausible stereoisomers of tris(DPAF-C$_9$) (**2**-C$_9$) via B3LYP/6-31G* level of theory using SPARTAN08 [34,35], the results revealed high stability of the *cis-cup*-form with other forms in stability order of *cup* > *chair* > *propeller*, as shown in Figure 1. This agreed well for the alkyl *n*-C$_6$, *n*-C$_7$, and *n*-C$_8$ substituents owing to the influence by strong dispersion interactions within the alkyl chains. In the case of the methyl and the ethyl substituents, *trans-chair*-form may have been more stable than *cis-cup*-form. Accordingly, three C$_4$-analogous substituents of tris(DPAF-C$_{2M}$) (**2**-C$_{2M}$) facing toward each other in the 3D molecular space above the central benzene ring should have brought in the minimum alkyl–alkyl interaction forces required to keep a slight favor of the *cis-cup*-form over either *trans-chair*- or *cis/trans-propeller* form.

Synthetically, the precursor molecule 2-bromo-9,9-bis(methoxyethyl)fluorene (BrF-C$_{2M}$) was prepared by alkylation of 2-bromofluorene with mesylated methoxyethanol reagent using potassium *t*-butoxide as a base in a high yield of 90% (Figure 1). The key 3D-conformeric intermediate **2**-C$_{2M}$ was synthesized by the reaction of BrF-C$_{2M}$ with 1,3,5-tris(phenylamino)benzene (TPAB) in the presence of sodium *t*-butoxide, a catalytic amount of tris(dibenzylideneacetone)dipalladium(0) [Pd$_2$(dba)$_3$(0)], and *rac*-2,2′-bis(diphenylphosphino)- 1,1′-binaphthyl (*rac*-BINAP, 0.75 mol%) in anhydrous toluene at refluxing temperature for a period of 72 h to yield 82% of the product as a light yellow solid after chromatographic purification [SiO$_2$, hexane–ethylacetate (1:1, *v/v*) as the eluent]. Subsequent attachment of three C$_{60}$(>DPAF-C$_9$) (**1**-C$_9$) on **2**-C$_{2M}$ should have led to the nanostructure of tris[(DPAF-C$_{2M}$)-C$_{60}$(>DPAF-C$_9$)] (**4**-C$_{2M-9}$) having all **1**-C$_9$ moieties extended outward from the central 1,3,5-triaminobenzene core. Prior to the attachment of three **1**-C$_9$ on **2**-C$_{2M}$, it was functionalized by the Friedel–Crafts acylation at C7 position of 2-diphenylaminofluorene moiety [36] with excessive α-bromoacetyl bromide (11.4 equiv.) in the presence of aluminum chloride (17 equiv.) in 1,2-dichloroethane at 0–10 °C to ambient temperature overnight to afford the corresponding α-bromoacetylfluorene derivative in this intermediate step of reactions. It resulted in viscous yellow semi-solids in 48% yield of tris(BrDPAF-C_{2M}) (3-C_{2M}). It was purified by either column or thin-layer chromatography (TLC) [silica gel, hexane–EtOAc (1:1, v/v) as eluent, R_f = 0.5 on TLC]. The final step of synthesis for the preparation of 3D-conformers 4-C_{2M-9} was performed by the reaction of 3-C_{2M} with 1-C_9 (5.0 equiv.) in the presence of 1,8-diazabicyclo[5.4.0]undec-7-ene (DBU) in toluene at room temperature for 8.0 h. During the first Bingel reaction period of 2.0 h, mono- and bis-adducts could be observed and detected by the TLC technique, showing two brown fraction bands (R_f = 0.2 and 0.3 on TLC plate, corresponding to mono- and bis-adducts, respectively). Subsequently, three brown fraction bands (R_f = 0.2, 0.3, and 0.4) appeared after 4 h of reaction, indicating sequential additions of C_{60}(>DPAF-C_9) on the starting substrate tris(BrDPAF-C_{2M}) (3-C_{2M}) with the R_f band at 0.2 gradually disappeared. At the end of the reaction, only two bands (R_f = 0.3 and 0.4) remained, with the latter being identified as a major product band of 4-C_{2M-9}. After purification of this fraction by column chromatography (silica gel) using toluene–ethyl acetate (7:3) as the eluent, the brown solids of tris[(DPAF-C_{2M})-C_{60}(>DPAF-C_9)] were obtained in 79% yield.

The compound 4-C_{2M-9} exhibited good solubility in common organic solvents owing to its possession of three DPAF-C_9 (with a total of six branched C_9-alkyl groups) and three DPAF-C_{2M} (with a total of six methoxyethyl groups) moieties (Figure 1), making three C_{60}> cages become encapsulated in the center of the 3D molecular configuration. The 3D configuration of *cis-cup* form resulted in these alkyl groups being the main structural moieties interacting with the solvent. Accordingly, the compound had solubility (20 mg/mL in $CHCl_3$ or CH_2Cl_2) over 10 times higher than that of C_{60} itself in toluene (1.4 mg/mL).

2.1. Spectroscopic Characterization of Synthetic 3D Configurated Fullerenyl Nanomaterials

All chemical conversions of intermediate chemicals to the corresponding products at each step of the reactions were characterized by various spectroscopic techniques. The functional attachment of three α-bromoaceto groups (3.0 equiv.) to tris(DPAF-C_{2M}) given the product of tris(BrDPAF-C_{2M}) was verified by both infrared (FT-IR) and ^{1}H-NMR spectra. The former showed a new strong carbonyl (–C=O) stretching absorption band centered at 1673 cm^{-1}, indicating each of the three carbonyl groups being bonded on a phenyl moiety, such as that of 2-C_{2M}. This absorption wavelength was in clear contrast to the strong band absorption at 1725 cm^{-1} normally detectable for an alkyl carbonyl group. In the case of ^{1}H-NMR spectrum, tris(BrDPAF-C_{2M}) displayed characteristic new peak signals of three methylene protons (H_α) next to the carbonyl group of the α-bromoaceto moiety at δ 4.49 (Figure 2Ab) as compared with that of 2-C_{2M} (Figure 2Aa). Subsequent attachment of three C_{60}(>DPAF-C_9) moieties to each of the three DPAF moieties of 3-C_{2M} with a cylopropylaceto bridging unit on each C_{60}> cage of three 1-C_9 (applied as a reagent) showed evidence of changing solubility characteristics of the product 4-C_{2M-9} matching with those of 1-C_9. Its FTIR spectrum displayed a slight shift of cyclopropyl keto group absorption band to v_{max} 1679 cm^{-1} (Figure 3d), which was assigned to the carbonyl (C=O) stretching band. It was also accompanied by an olefinic (C=C) absorption band centered at v_{max} 1591 cm^{-1}. Both C=O and C=C bands were correlated to those of C_{60}(>DPAF-C_9) (Figure 3b) and tris[C_{60}(>DPAF-C_9)] (Figure 3c), showing a nearly identical absorption wavenumber. Most importantly, we were able to detect two typical fullerenyl cage bands at v_{max} 574 (w) and 524 (s) cm^{-1} (Figure 3d). These two bands were corresponding characteristic absorptions used to provide evidence of (C_{60}>)-related monoadducts and bisadducts with absorption wavenumbers and relative intensity ratios differentiable from those of C_{60} (Figure 3a) and 1-C_9 substituents (Figure 3b). Accordingly, we applied this IR technique for the product structure verification during the chemical conversion from 3-C_{2M} to 4-C_{2M-9}. Upon conversion of C_{60} to its monoadducts, such as those of Figure 3b,c, the remaining cage structure of C_{60}> exhibited the same two bands with a reduced peak intensity for the 574 cm^{-1} band. The intensity of this band was further reduced in the structure of

<C$_{60}$>-like bisadduct, such as **4**-C$_{2M-9}$ (Figure 3d). Furthermore, the latter band at 524 cm^{-1} still remaining strong was indicative of successful attachment of C$_{60}$(>DPAF-C$_9$) moieties on **3**-C$_{2M}$ due to the possibility of having the second malonate bridging unit being attached at or near the equator region of the C$_{60}$> cage. This would have led to the retention of a C$_{60}$ half-cage that enabled absorption at 524 cm^{-1}.

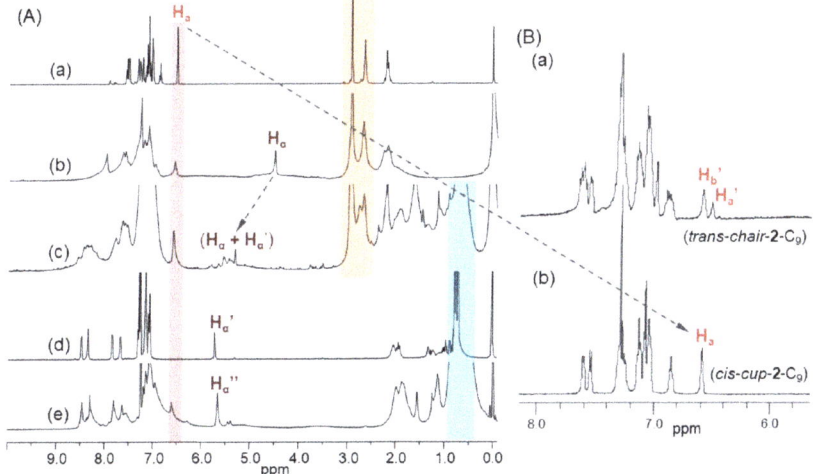

Figure 2. ^{1}H-NMR spectra (CDCl$_3$) of (**A**) (a) tris(DPAF-C$_{2M}$) (**2**-C$_{2M}$), (b) tris(BrDPAF-C$_{2M}$) (**3**-C$_{2M}$), (c) tris[(DPAF-C$_{2M}$)-C$_{60}$(>DPAF-C$_9$)] (**4**-C$_{2M-9}$), (d) C$_{60}$(>DPAF-C$_9$) (**1**-C$_9$), and (e) tris[C$_{60}$(>DPAF-C$_9$)]. (**B**) ^{1}H-NMR spectra (CDCl$_3$) of (a) *trans-chair*-tris(DPAF-C$_9$) (*trans-chair*-**2**-C$_9$) and (b) *cis-cup*-tris(DPAF-C$_9$) (*cis-cup*-**2**-C$_9$) for comparison.

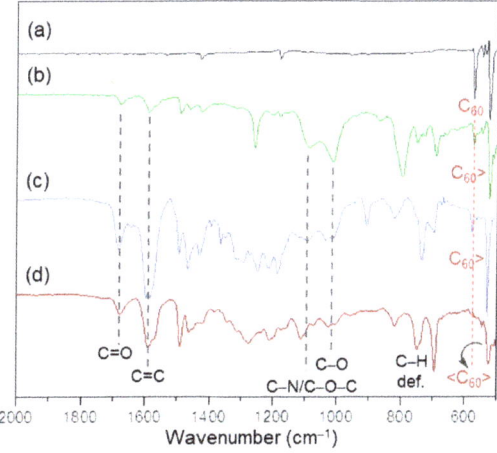

Figure 3. Infrared spectra (KBr) of (a) C$_{60}$, (b) C$_{60}$(>DPAF-C$_9$) (**1**-C$_9$), (c) tris[C$_{60}$(>DPAF-C$_9$)]. and (d) tris[(DPAF-C$_{2M}$)-C$_{60}$(>DPAF-C$_9$)] (**4**-C$_{2M-9}$).

By the analysis of ^{1}H-NMR spectrum, disappearance of peaks corresponding to the chemical shift of α-proton (H$_\alpha$) on the α-bromoaceto group of **3**-C$_{2M}$ at δ 4.49 (Figure 2Ab) along with the appearance of new peaks over δ 5.25–5.78 (Figure 2Ac) provided evidence of successful formation of a cyclopropanyl keto-bridging unit between a C$_{60}$> cage and the fluorene moiety. By using the previously reported chemical shift values of the α-proton ($H_\alpha{}'$) in $C_{60}(>DPAF-C_9)$ (Figure 2Ad) [20] and the related α-proton ($H_\alpha{}''$) in tris(DPAF-C$_9$) (2-C$_9$, Figure 2Ae) [23] as the reference, we assigned these proton peaks to a combination of H_α and $H_\alpha{}'$. A large downfield shift of the H_α chemical shift from those of 3-C$_{2M}$ at δ 4.49 to δ 5.25–5.78 for 4-C$_{2M-9}$ provided clear evidence of three $C_{60}(>DPAF-C_9)$ moieties being attached on the corresponding α-bromoaceto bridging units of 3-C$_{2M}$. The characteristics of the multipeaks for H_α and $H_\alpha{}'$ revealed a less symmetrical environment among these six protons of 4-C$_{2M-9}$ as the geometric shape of the nanostructure extended to a 3D configuration. It is worthwhile to mention that a large down-fielded chemical shift value of either $H_\alpha{}'$ or $H_\alpha{}''$ away from the normal value of δ 2.1–2.5 for an alkyl aceto-α-proton was caused by the influence of strong [60]fullerenyl ring current in close vicinity. In addition, the alkyl proton regions over δ 2.93 (methoxy proton, 18H) and δ 2.75–2.64 (ethylenoxy proton, 12H) of Figure 2Ac (marked by beige) were correlated well to those of 2-C$_{2M}$ at δ 2.91 (18H) and δ 2.64 (12H) (Figure 2Aa), respectively, indicating good retention of central tris(DPAF-C$_{2M}$) core region without any structural change of methoxy groups during the Friedel–Crafts acylation reaction. It also showed a new group of methyl proton peaks at δ 0.30–2.07 having an integration ratio value of 113.28, which represented 114 fluorenyl protons of C$_9$-alkyl proton (19Hs for each of the two C$_9$-alkyls of DPAF-C$_9$) and was indicative of six C$_9$-alkyl groups in the structure, consistent with the product structure. Additional ^{1}H-NMR spectroscopic data analyses on proton integrations of all proton peaks to substantiate and count for the molar quantity ratio among fluorene, methoxyethyl, and C$_9$ alkyl moieties to prove the molecular formulation of 4-C$_{2M-9}$ are provided in supporting information.

Most importantly, characteristics of central benzene protons at the core region could be used for the analysis of the relatively geometric configuration of three fluorenyl rings with respect to each other. With a symmetrical structure of TPAB, three benzene protons should have displayed a singlet peak in its ^{1}H-NMR spectrum. Upon attachment of a bulky fluorenyl moiety at each diphenylamino group, it induced high torsional stress and steric hindrance at the nitrogen atom that forced each 9,9'-di(methoxyethyl)fluorene moiety to twist or rotate either upward or downward from the central benzene plane. The action resulted in two main 3D conformers: cis-cup-2-C$_{2M}$ and trans-chair-2-C$_{2M}$. The former with three $C_{60}(>DPAF-C_9)$ moieties facing upward on the same side in the structure gave a singlet H_a peak (Figure 1). The latter with one facing downward and two $C_{60}(>DPAF-C_9)$ moieties facing upward in the structure resulted in two proton peaks for trans-$H_a{}'$ (1H) and trans-$H_b{}'$ (2H) (Figure 2Ba). By analyzing Figure 2Aa of tris(DPAF-C$_{2M}$), a sharp singlet proton peak at δ 6.53 was assigned to the chemical shift of central benzene proton cis-H_a. This peak was compared with that of the H_a proton peak of cis-cup-tris(DPAF-C$_9$) (cis-cup-2-C$_9$, Figure 2Bb) showing even better resolution of the peak profile, indicating a high purity of one 3D conformer fraction in a cis-cup-2-C$_{2M}$ form. Surprisingly, this conformer fraction was, in fact, the major product. Apparently, the hydrophobic–hydrophobic dispersion interaction forces derived from three methoxyethyl chains and heteroatoms were stronger than those among all C$_4$-alkyl groups, which led to higher tendency in formation of the cis-cup-form. Accordingly, subsequent attachment of three $C_{60}(>DPAF-C_9)$ moieties on cis-cup-2-C$_{2M}$ led to a similar formation of corresponding cis-cup-tris[(DPAF-C$_{2M}$)-$C_{60}(>DPAF-C_9)$] (cis-cup-4-C$_{2M-9}$), all having 1-C$_9$ moieties facing upward from the central benzene core at the same side with respect to each other. Additional structural analyses and discussions are provided in supporting information.

In the case of the potential formation of regio-isomers of 4-C$_{2M-9}$ at the <C$_{60}$> moiety, since the monoadduct structure of $C_{60}(>DPAF-C_9)$ was well-defined, with the assistance of X-ray single crystal structural analysis of $C_{60}(>DPAF-C_2)$ [31,36], its attachment on tris(BrDPAF-C$_{2M}$) (3-C$_{2M}$) was believed to be governed by the bulkiness and the steric hindrance of both relatively large entities to result in only a limited number of region isomers on the C$_{60}$ cage. This was proven by the ^{1}H-NMR spectrum of 4-C$_{2M-9}$ showing only several H_α and $H_\alpha{}'$ proton peaks at roughly δ 5.2–5.8 (Figure 2Ac) instead of the broad band normally seen for the existence of a large number of region isomers. To our surprise, a peak at 524 cm^{-1} assigned the characteristic infrared absorption band of a half-C$_{60}$ cage (as stated above) showed close resemblance to those of the monoadduct $C_{60}(>DPAF-C_9)$ (Figure 3b) and tris[C_{60}(>DPAF-C_9)] (Figure 3c) at an identical wavenumber. This implied the structure of the major regio-isomeric products had both addend moieties located at the same half-sphere of a C_{60} cage that left the other half-sphere of C_{60} untouched.

2.2. Photophysical and Physical Properties of 3D Conformeric Fullerenyl Nanomaterials

Photophysical properties of the 3D conformer *cis-cup*-**4**-C_{2M-9} were compared with those of precursor intermediates using the UV-vis spectroscopic technique. They were governed by two photoresponsive moieties, including three electron (e^-)-accepting fullerene cages and six light-harvesting DPAF antenna units as electron (e^-)-donors. The use of the latter was to enhance the optical absorption capability at longer visible wavelengths. The absorption wavelength could be varied and modulated by the appropriate chemical modification of functional substituents on fluorenyl moiety to affect electron-pushing (donating) and pulling (accepting toward the molecular edge of C_{60}> cage moiety) mobility across the molecular π-conjugation system. As shown in Figure 4Ae of *cis-cup*-**4**-C_{2M-9}, optical absorption of C_{60}> cage moieties appeared mainly at the broad band centered at 296 nm (1.82×10^5 L mol^{-1} cm^{-1}), whereas the band centered at 411 nm (1.11×10^5 L mol^{-1} cm^{-1}) was attributed to the absorption of DPAF moieties. Characteristics of the latter band were compared with those of tris(DPAF-C_{2M}) (Figure 4Aa), tris(BrDPAF-C_{2M}) (Figure 4Ab), *cis-cup*-tris[C_{60}(>DPAF-C_9)] (Figure 4Ac), and C_{60}(>DPAF-C_9) (**1**-C_9, Figure 4Ad), showing a clear bathochromic shift of the 351 nm band of **2**-C_{2M} to 406 nm (7.91×10^4 L mol^{-1} cm^{-1}) of **3**-C_{2M}, which matched roughly with the 404 nm band of **1**-C_9 and the 402 nm band *cis-cup*-tris[C_{60}(>DPAF-C_9)] for the peak assignment. This assignment was also consistent with the observation of roughly equal absorption extinction coefficient (ε) values for **2**-C_{2M}, **3**-C_{2M}, and tris[C_{60}(>DPAF-C_9)] with the same three DPAF moieties per molecule. Upon the attachment of three C_{60}(>DPAF-C_9) moieties, optical absorptions of [60]fullerene moieties of *cis-cup*-**4**-C_{2M-9} (Figure 4Ae) at 296 nm became dominant in the spectrum with a higher ε value. It was accompanied by a weak characteristic (forbidden) steady state absorption band of the C_{60}> moiety appearing at 692 nm (the insert of Figure 4A) with a slightly higher extinction coefficient for the monoadduct **1**-C_9 than the bisadduct *cis-cup*-**4**-C_{2M-9}, which was also consistent with the photophysical property discussion above and provided further confirmation of a conjugated fullerenyl nanostructure.

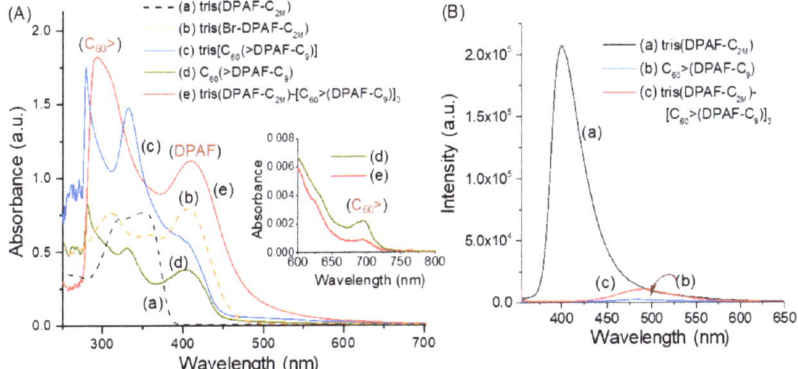

Figure 4. UV-vis spectra of (**A**) (a) tris(DPAF-C_{2M}) (**2**-C_{2M}), (b) tris(BrDPAF-C_{2M}) (**3**-C_{2M}), (c) *cis-cup*-tris[C_{60}(>DPAF-C_9)], (d) C_{60}(>DPAF-C_9) (**1**-C_9), and (e) *cis-cup*-**4**-C_{2M-9}, where (a), (b), and (e) were taken in CDCl$_3$ and (c) and (d) were taken in toluene. (**B**) Fluorescence spectra of (a) tris(DPAF-C_{2M}) (λ_{ex}: 352 nm), (b) C_{60}(>DPAF-C_9) (λ_{ex}: 406 nm), and (c) *cis-cup*-**4**-C_{2M-9} (λ_{ex}: 410 nm). The concentration of all samples is 1.0×10^{-5} M.

In addition, a roughly 2.1-fold higher ε value (1.11×10^5 L mol^{-1} cm^{-1}) of the 411 nm peak in Figure 4Ae compared to that of 404 nm band of *cis-cup*-tris[C_{60}(>DPAF-C_9)] was consistent with a double number of DPAF arms per molecule for the former. Furthermore, very efficient intramolecular energy transfer from the excited singlet state of DPAF-C_9 antenna to C_{60}> was detected, which nearly eliminated the fluorescence of C_{60}(>DPAF-C_9) (λ_{ex}: 406 nm, Figure 4Bb). In high contrast, without any C_{60}> cage in the structure, the compound of tris(DPAF-C_{2M}) showed a strong intensity of fluorescence emission (λ_{ex}: 352 nm, Figure 4Ba) that clearly indicated the loss of photoexcited energy being associated with the influence of [60]fullerene. With an additional DPAF-C_9 antenna in the structure of 4-C_{2M-9}, it began to experience a slightly excessive fluorescence emission (λ_{ex}: 410 nm, Figure 4Bc) after the majority of photoexcited DPAF-C_{2M} energy underwent direct intramolecular energy transfer to the closely bonded [60]fullerene cage.

In investigating the plausibility of photoinduced intramolecular electron (e$^-$)-transfer capability within the nanostructure of the 3D conformer cis-cup-tris[(DPAF-C_{2M})-C_{60}(>DPAF-C_9)] (cis-cup-4-C_{2M-9}), we first investigated the unit character of redox potentials among all structural components, including the bisadduct-based <C_{60}> cage and the DPAF-C_9 moieties for comparison using the cyclic voltammetric (CV) technique. Several CV measurements were performed on the sample of cis-cup-4-C_{2M-9} in a solution of CH_2Cl_2 containing (n-butyl)$_4$N$^+$-PF_6^- as the electrolyte and Pt as both the working and the counter electrodes and with Ag/AgCl as the reference electrode.

To deliver appropriate redox potential analyses and data interpretation, related CV characteristics of a C_{60}-bisadduct of C_{60}(>t-Bu-malonate)$_2$ with a <C_{60}> cage attached by two t-butylmalonate groups and the precursor compound C_{60}(>DPAF-C_9) were collected. They were performed under the same CV condition over cyclic oxidation and reduction voltages versus Ag/Ag$^+$ from −2.0 to 2.0 V as those for cis-cup-4-C_{2M-9}, as shown in Figure 5. As a result, it displayed one reversible oxidation ($^1E_{ox}$ of 1.51 V) reduction ($^1E_{red}$ of 1.32 V) cyclic wave with the first half wave oxidation potential ($^1E_{1/2,ox}$) of 1.42 V for the DPAF moieties of cis-cup-4-C_{2M-9} at positive voltages (Figure 5Ab,Bb). In the negative voltage region, its CV diagram displayed three reversible reductions at −0.34 ($^1E_{red}$), −0.82 ($^2E_{red}$), and −1.29 V ($^3E_{red}$) with the corresponding cyclic oxidation waves at −0.15 ($^1E_{ox}$), −0.44 ($^2E_{ox}$), and −0.96 ($^3E_{ox}$), respectively. These data corresponded to the first to the third half wave reduction potentials of −0.25 ($^1E_{1/2,red}$), −0.63 ($^2E_{1/2,red}$), and −1.12 V ($^3E_{1/2,red}$), respectively. By comparison of these values to those of C_{60}(>t-Bu-malonate)$_2$ (Figure 5Aa,Ba) for the fullerene cage moiety and those of C_{60}(>DPAF-C_9) (Figure 5Ac,Bc) for both DPAF and fullerene cage moieties, highly consistent and reproducible redox potential characteristics among structural components were found that also substantiated the structural derivatization of tris[(BrDPAF-C_{2M}) (3-C_{2M}) with triple C_{60}(>DPAF-C_9) to form cis-cup-4-C_{2M-9}. Accordingly, the latter exhibited combined CV characteristics of <C_{60}> and DPAF-C_9. These CV characteristics were reproducible for four repeated redox cycles with the reductive C_{60}> and the oxidative DPAF potential profiles showing only slight changes at the potential range of −2.0 to 2.0 V. This implied good stability of the material under CV conditions that led to possible reuse of 4-C_{2M-9}.

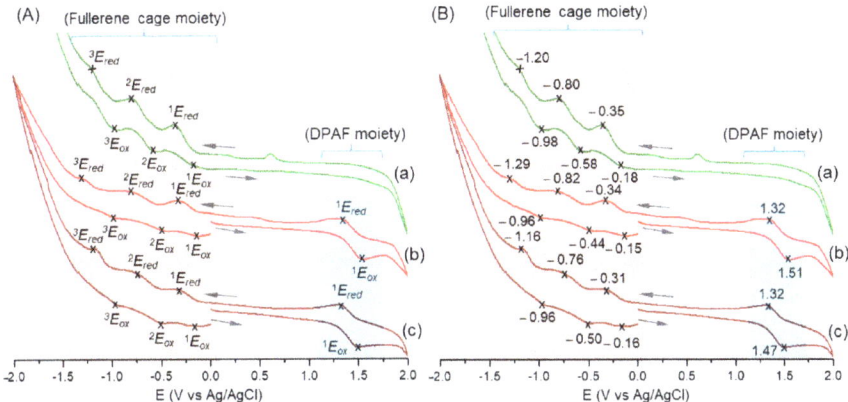

Figure 5. Cyclic voltammograms (CV) of (a) $C_{60}(>t\text{-Bu-malonate})_2$, (b) *cis-cup*-tris[(DPAF-C_{2M})-$C_{60}(>$DPAF-$C_9)$], and (c) $C_{60}(>$DPAF-$C_9)$, where (**A**) displays the sequential redox cycles of each compound with the assignments and (**B**) shows the corresponding potential voltage values at either the peak maximum or minimum of each redox cycle. All solutions were in a concentration of 5.0×10^{-3} M in CH_2Cl_2 using (n-butyl)$_4$N$^+$-PF$_6^-$ as the electrolyte (0.1 M), Pt as working and counter electrodes, and Ag/AgCl as the reference electrode at a scan rate of 10 mV/s.

2.3. Evidence of Intramolecular Energy- and Electron-Transfer Events within cis-cup-4-C_{2M-9} by Detection of Corresponding Reactive Oxygen Species (ROS)

There is an appropriate approach to substantiate intramolecular energy and electron transfer events within the nanomolecular structure of *cis-cup*-4-C_{2M-9} by directly detecting the photoinduced production of reactive oxygen species (ROS). In general, the most common ROS includes singlet oxygen ($^1O_2$) produced by the Type-II photomechanism via the intermolecular transfer of triplet energy to molecular oxygen (O_2) and superoxide radical ($O_2^-\cdot$) generated by the intermolecular transfer of electron (e$^-$) to O_2. For the former case, upon photoexcitation at the $C_{60}>$ cage moiety of *cis-cup*-4-C_{2M-9}, the singlet excited state of bis-methanofullerenyl $^1(<C_{60}>)^*$ may undergo facile intersystem crossing in a nearly quantitative efficiency to its triplet excited state $^3(<C_{60}>)^*$ that can be accounted for by the efficient production of $^1O_2$ via Type-II triplet energy transfer processes. Alternatively, if the photoexcitation process is aimed at either the DPAF-C_{2M} or the DPAF-C_9 moiety of *cis-cup*-4-C_{2M-9}, the resulting corresponding singlet excited states of either $^1(\text{DPAF-}C_{2M})^*$ or $^1(\text{DPAF-}C_9)^*$ may undergo both pathways of (1) intramolecular energy transfer from either $^1(\text{DPAF-}C_{2M})^*$ or $^1(\text{DPAF-}C_9)^*$ to the $<C_{60}>$ moiety to produce neutral DPAF-C_{2M} or DPAF-C_9 and $^1(<C_{60}>)^*$, respectively; (2) intramolecular electron (e$^-$)-transfer from either $^1(\text{DPAF-}C_{2M})^*$ or $^1(\text{DPAF-}C_9)^*$ to the $<C_{60}>$ moiety to produce cationic (DPAF-C_{2M})$^{+\cdot}$ or (DPAF-C_9)$^{+\cdot}$ and $(<C_{60}>)^{-\cdot}$, respectively. Both events of (1) and (2) can occur concurrently. Subsequent intermolecular e$^-$-transfer from $(<C_{60}>)^-$ to O_2 produces the corresponding neutral $<C_{60}>$ and O_2^- following the Type-I photomechanism.

Accordingly, by the direct detection of ROS on either $^1O_2$ and/or $O_2^-\cdot$ upon irradiation on *cis-cup*-4-C_{2M-9} at either $<C_{60}>$ or DPAF-C_{2M}/DPAF-C_9 moiety, we were able to provide the evidence of intramolecular energy and electron transfer processes happening within this 3D-conformer. We selected two reliable fluorescent (FL) probes for the detection of either $^1O_2$ or $O_2^-\cdot$ separately in the solution of *cis-cup*-4-C_{2M-9} with high selectivity and specificity as a crucial measure. To detect the former ROS $^1O_2$, a synthetic highly fluorescent compound α,α'-(anthracene-9,10-diyl)-bis(methylmalonic acid) (ABMA) was used as the probe in the experiment. Its UV-vis absorption and fluorescence emission spectra are given in Figure 6a,b, respectively. In the probe reaction, chemical trapping of $^1O_2$ by highly fluorescent ABMA resulted in the formation of non-fluorescent 9,10-endoperoxide product ABMA-O_2 (Figure 7A). This chemical conversion allowed us to follow the intensity loss of fluoresce emission upon photoexcitation. The loss could be associated with the proportional quantity of $^1O_2$ produced. The correlation was valid owing to a higher reaction kinetic rate of the trapping process in solution than the internal decay of $^1O_2$ in the same solvent system of a DMF–CHCl$_3$ (1:9, v/v) mixture. Experimentally, the quantity of $^1O_2$ generated was monitored and counted by the relative intensity decrease of fluorescence emission (λ_{em}) of ABMA at 428 nm under excitation wavelength (λ_{ex}) of 380 nm. At this excitation wavelength, it matched partially with the optical absorption band of DPAF moieties of cis-cup-4-$C_{2M\text{-}9}$ that led to a slight fluorescence emission (Figure 6c) after the intramolecular energy and the e$^-$-transfer processes. It gave a slightly higher count in the overall FL intensity during the experiment (Figure 6d). In a typical probe reaction, a master solution of ABMA in DMF was diluted by CHCl$_3$ prior to the addition of cis-cup-4-$C_{2M\text{-}9}$. It was followed by periodical illumination using a light emitting diode (LED) lamp of white light (a power output of >2.0 W) operated at two major emission peak maxima (λ_{max}) centered at 451 and 530 nm. The former light emission spectrum exhibited a sufficient bandwidth covering the 410–470-nm region for photoexcitation of DPAF moieties with optical absorption bands covering 380–500-nm (Figure 4Ae). As a result, we were able to detect rapid production of $^1O_2$ by cis-cup-4-$C_{2M\text{-}9}$ upon irradiation in a decreasing curve profile over a period of more than 120 min (Figure 7Ab).

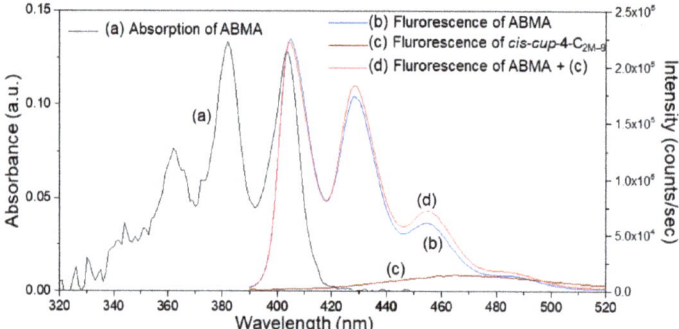

Figure 6. (a) UV-vis spectra of α,α'-(anthracene-9,10-diyl)bis(methylmalonic acid) (ABMA) and fluorescence (FL) emission spectrum of (b) ABMA, (c) cis-cup-tris[(DPAF-C_{2M})-C_{60}(>DPAF-C_9)] (cis-cup-4-$C_{2M\text{-}9}$), and (c) a combination of ABMA and cis-cup-4-$C_{2M\text{-}9}$ in a solvent mixture of DMF–CHCl$_3$ (1:9, v/v) in a concentration of 10^{-6} M.

The FL probe experiments were calibrated by a blank control run using the same probe concentration of ABMA alone and an illumination time scale in the absence of cis-cup-4-$C_{2M\text{-}9}$ (Figure 7Aa). Apparently, we observed slight photodegradation of ABMA itself. This may have implied the existence of a photoinduced triplet state of ABMA in a low quantity due to exposure to short wavelength regions of the light emission bandwidth covering over ~380 nm of ABMA absorption bands.

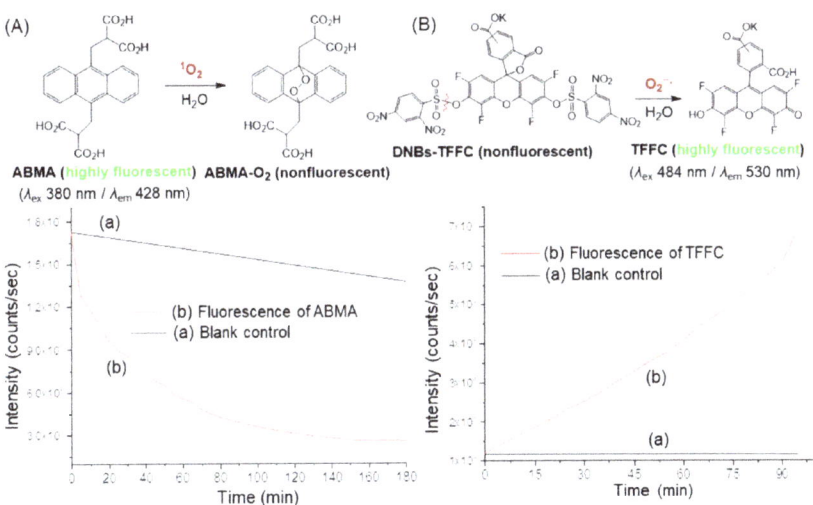

Figure 7. Fluorescence (FL) emission spectra of (**A**) ABMA and (**B**) tetrafluorofluorescein-10′ (or 11′)-carboxylate (TFFC) to correlate directly the singlet oxygen ($^1O_2$) and superoxide radical ($O_2^{-}\cdot$) production efficiency, respectively, with (a) blank control and (b) a mixture of corresponding probe and cis-cup-**4**-C$_{2M-9}$ samples in DMF–CHCl$_3$ (1:9, v/v) at a concentration of 1.0×10^{-5} M using ABMA as the $^1O_2$ trapping agent at λ_{ex} 350 nm and λ_{em} 428 nm and DNBs-TFFC as the $O_2^{-}\cdot$-acceptor agent at λ_{ex} 484 nm and λ_{em} 530 nm for detection with the irradiation of white light emitting diode (LED) light.

In the case of detecting superoxide radical as the second ROS, a synthetic $O_2^{-}\cdot$-reactive fluorescent probe precursor molecule, non-fluorescent potassium bis(2,4-dinitrobenzenesulfonyl)-2′,4′,5′,7′-tetrafluorofluorescein-10′ (or 11′)-carboxylate (DNBs-TFFC), was applied for the experiment. Its molecular structure was synthetically modified from that reported previously [38], showing good reaction selectivity with a high $O_2^{-}\cdot/^1O_2$ sensitivity ratio. Since DNBs-TFFC itself is photodegradable, a dialysis film with the molecular weight cut-off (MWCO) of 100–500 Daltons was applied to hold the solution of cis-cup-**4**-C$_{2M-9}$ in toluene–DMSO (9:1, v/v). The sack bag was separated from the solution of the probe DNBs-TFFC. The latter was kept in a cuvette with stirring during the fluorescence emission measurement. Only the solution of cis-cup-**4**-C$_{2M-9}$ in the dialysis membrane sack was subjected to the white LED light exposure. Any superoxide radical produced was allowed to rapidly diffuse into the probe solution through the dialysis membrane and initiate the desulfonylation of DNBs-TFFC. The $O_2^{-}\cdot$-trapping reaction led to the elimination of two dinitrobenzenesulfonyl moieties and yielded the corresponding bisphenol intermediate, as shown in Figure 7B. Rearrangement of the bisphenol intermediate to the ring-opening of lactone afforded highly fluorescent potassium 2′,4′,5′,7′-tetrafluorofluorescein-10′ (or 11′)-carboxylate regioisomers (TFFC). The latter compound gave the fluorescence emission at 530 nm (λ_{em}) with the excitation at 484 nm (λ_{ex}). As the probe DNBs-TFFC was not a fluorescent compound, detected emission photon counts were fully associated with the quantity of TFFC produced. Measured total emission intensity counts were then correlated to the relative quantity of $O_2^{-}\cdot$ generated. As shown in Figure 7Bb, nearly linear progressive increase of fluorescence intensity counts over the full irradiation period was observed that revealed the constant production of $O_2^{-}\cdot$ from the photoexcited cis-cup-**4**-C$_{2M-9}$. As discussed above, continuous irradiation on six DPAF moieties of cis-cup-**4**-C$_{2M-9}$ by white LED light (2.0 W) stimulated photoexcitation from the ground to the singlet excited state. Subsequent intramolecular e^{-}-transfer from ^{1}(DPAF-C$_n$)* to <C$_{60}$> moieties resulted in the formation of anionic [60]fullerenyl bisadduct radical (<C$_{60}$>)$^{-}$ intermediate. In the presence of O_2, it was followed by further e^{-}-transfer from (<C$_{60}$>)$^{-}$ intermediate to O_2 to produce O_2^- in a sequential multiple-step Type-I photomechanism. These results clearly provided the evidence of photoinduced intramolecular e^--transfer mechanism within the 3D-conformer cis-cup-**4**-C_{2M-9}.

Furthermore, one of the main crucial criterion for the use of these types of C_{60}-(light-harvesting antenna)$_n$ conjugates, such as **4**-C_{2M-9}, as the nano-photosensitizers for antibacterial inactivation (aPDI) is their high photostability. Unlike the conventional organic chromophore-based photosensitizers suffering rapid photodegradation, C_{60}-(light-harvesting antenna)$_n$ based nano-drugs were found to be capable of a single dose with multiple aPDI/PDT (photodynamic therapy) treatments [39–42].

3. Experimental Section

3.1. Chemicals and Reagents

Reagents of sodium *t*-butoxide, α-bromoacetyl bromide, aluminum chloride ($AlCl_3$), 1,8-diazabicyclo[5.4.0]undec-7-ene (DBU), *rac*-2,2'-bis(diphenylphosphino)-1,1'-binaphthyl (BINAP), and tris(dibenzylideneacetone)dipalladium(0) [$Pd_2(dba)_3(0)$] were purchased from Aldrich Chemicals, Natick, Massachusetts, USA and used without further purification. Fullerene materials with a purity of 99% were purchased from Suzhou Dade Carbon Nanotechnology Co., Ltd. Suzhou, Jiangsu, China. Anhydrous grade solvent of tetrahydrofuran (THF) was used and further dried via refluxing over sodium and benzophenone overnight and distilled under reduced pressure (10^{-1} mmHg). The precursor compounds including 1,3,5-tris(*N*-phenylamino)benzene (TPAB) and 2-bromo-9,9-di(methoxyethyl)fluorene (BrF-C_{2M}) were synthesized according to our previous procedures [34].

3.2. Instruments for Spectroscopic Measurements

^{1}H-NMR spectra were recorded on either Bruker and Spectrospin Avance 500 or Bruker AC-300 spectrometer. UV-vis spectra were recorded on a PerkinElmer *Lambda 750* UV spectrometer. Fluorometric traces were collected using a PTI QuantaMaster™40 Fluorescence Spectrofluorometer. The light source used in this experiment included a collimated white LED light with an output power of 2.0 W (Prizmatix, Southfield, MI, USA). Infrared spectra were recorded as KBr pellets on a Thermo Nicolet *AVATAR 370* FTIR spectrometer. Cyclic voltammetry (CV) was record on EG&G Princeton Applied Research 263A Potentiostat/Galvanostat using Pt metal as the working electrode, Ag/AgCl as the reference electrode, and Pt wire as the counter electrode at a scan rate of 10 mV/s. The solution for CV measurements was prepared in a concentration of 1.0–5.0 × 10^{-3} M in appropriate solvents containing the electrolyte Bu_4N^+-PF_6^- (0.1 M).

3.3. Synthesis of N^1,N^3,N^5-Tris(9,9-di(methoxyethyl)fluoren-2-yl)-1",3",5"-tris(phenylamino)-benzene as Tris(DPAF-C_{2M}) (**2**-C_{2M})

Synthetic procedure for the preparation of tris(DPAF-C_{2M}) was slightly modified from those methods reported recently [34]. In general, a mixture of BrF-C_{2M} (7.33 g, 20.3 mmol, excess), TPAB (1.16 g, 3.3 mmol), and sodium *t*-butoxide (1.94 g, 20.3 mmol) was dissolved in anhydrous toluene (75 mL) and stirred for 1 h to give a homogeneous solution. The catalyst $Pd_2(dba)_3(0)$ (0.023 g, 0.25 mol%) and *rac*-BINAP (0.046 g, 0.75 mol%) were added to the solution, followed by heating to refluxing temperature under nitrogen for a period of 72 h. After cooling the resulting mixture to room temperature, it was washed with water three times by extraction, the organic layer was separated, and it was dried over sodium sulfate. After solvent evaporation, a small quantity of crude paste was tested on the TLC plate to show the major product at R_f = 0.6 using hexane–ethylacetate (1:1, *v/v*) as the eluent. This product spot had a dense yellow-brown color on the top accompanied by a light visible tail. The tail portion was assumed to be the product in the *trans-chair* form. This tail portion was subsequently separated from the main top portion of the *cis-cup* form via column chromatography, followed by the TLC plate purification using silica gel as the stationary phase and hexane-ethylacetate

(1:1, v/v) as the eluent to afford the product of cis-cup-tris(DPAF-C$_{2M}$) (2-C$_{2M}$) as light yellow solids in 82% yield (3.23 g). Verification of the cis-cup form was based on the detection of a singlet proton peak at δ6.53 (H$_a$) corresponding to three central core 1,3,5-benzene protons, whereas the trans-chair form gave two proton groups at a integration ratio of 2:1. Spectroscopic data: FT-IR (KBr) v_{max} 3062 (w, aromatic C-H stretching), 3031 (w), 3016 (w), 2969 (w, aliphatic C-H stretching), 2922 (m), 2875 (m), 2814 (w), 1571 (s, C=C), 1491 (s, anti-symmetric deformations of CH$_3$ groups and scissor vibrations of CH$_2$ groups), 1448 (s), 1428 (m), 1381 (w, symmetric deformations of CH$_3$ groups), 1292 (m, asymmetric stretching vibrations of C-N-C), 1248 (m, asymmetric stretching vibrations of C-N-C), 1212 (w), 1176 (w), 1155 (w), 1110 (s, stretching vibrations of C-O-C), 1039 (w), 948 (w), 877 (w), 829 (w), 754 (m), 738 (s, out-of-plan deformation of C-H), 711 (m, out-of-plan deformation of C-H), 692 (s), and 511 (w) cm^{-1}; UV-vis (CHCl$_3$, 1.0×10^{-5} M) λ_{max} (ε) 321 (7.24×10^4 L mol^{-1} cm^{-1}) and 351 nm (7.66×10^4 L mol^{-1} cm^{-1}); ^{1}H NMR (500 MHz, CDCl$_3$) δ 7.58 (s, 3H, br), 7.52 (d, 3H), 7.35–7.24 (m, 9H), 7.15–7.03 (m, 18H), 6.89 (t, 3H), 6.53 (s, 3H, central benzene protons H$_a$), 2.91 (s, 18H), 2.65 (m,12H), and 2.19 (m, 12H).

3.4. Synthesis of N^1,N^3,N^5-Tris(7-α-bromoacetyl-9,9-di(methoxyethyl)fluoren-2-yl)-1",3",5"-tris-(phenylamino)benzene as Tris(BrDPAF-C$_{2M}$) (3-C$_{2M}$)

The compound cis-cup-tris(DPAF-C$_{2M}$) (cis-cup-2-C$_{2M}$, 0.53 g, 0.44 mmol) was added to a homogeneous suspension of AlCl$_3$ (1.0 g, 7.5 mmol) in 1,2-dichloroethane (40 mL) at 0 °C with vigorous stirring. The reagent α-bromoacetyl bromide (1.0 g, 5.0 mmol) was added slowly over 10 min while maintaining the temperature between 0–10 °C. The mixture was then stirred overnight at room temperature. An excessive amount of AlCl$_3$ remaining in the solution was quenched by slow addition of water (50 mL) while maintaining the temperature below 45 °C. After washing sequentially with dil. HCl (1.0 N, 50 mL) and water (50 mL × 2), the organic layer was separated and dried over sodium sulfate and then concentrated in vacuo to give the crude product as viscous yellow semi-solids. It was purified by column chromatography (silica gel) followed by thin-layer chromatography (TLC) using hexane–EtOAc (1:1, v/v) as eluent to afford cis-cup-tris(Br-DPAF-C$_{2M}$) (cis-cup-3-C$_{2M}$) (at R_f = 0.5 on TLC plate) in 48% yield (0.33 g). Spectroscopic data: FT-IR (KBr) v_{max} 3062 (w, aromatic C-H stretching), 3031 (w), 3016 (w), 2969 (w, aliphatic C-H stretching), 2922 (m), 2875 (m), 1673 (s, C=O), 1571 (s, C=C), 1491 (s, anti-symmetric deformations of CH$_3$ groups and scissor vibrations of CH$_2$ groups), 1467 (s), 1448 (s), 1428 (m), 1388 (w), 1348 (w), 1282 (s), 1251 (s), 1195 (m), 1176 (w), 1110 (s, stretching vibrations of C-O-C), 1035 (w), 948 (w), 879 (w), 823 (m), 755 (m), 740 (s, C-H out-of-plan deformation), 715 (m, C-H out-of-plan deformation), 694 (s), 617 (w) and 538 (w) cm^{-1}; UV-vis (CHCl$_3$, 1.0×10^{-5} M) λ_{max} (ε) 310 (7.65×10^4 L mol^{-1} cm^{-1}) and 406 nm (7.91×10^4 L mol^{-1} cm^{-1}); ^{1}H NMR (500 MHz, CDCl$_3$) δ 7.98 (m, 3H), 7.58 (m, 3H), 7.34–7.20 (m, 9H), 7.11–6.98 (m, 18H), 6.96 (m, 3H), 6.56 (m, 3H, aromatic protons of central phenyl ring), 4.49 (m, 6H, α-proton next on C$_{61}$), 2.91 (s, 18H. primary C$_{2M}$ alkyl protons next to O-atom), 2.81–2.50 (centered at δ 2.66, m, 12H, secondary C$_{2M}$ alkyl protons next to O-atom), 2.48–2.07 (centered at δ 2.16, m, 12H, C$_{2M}$ alkyl protons next to the fluorene ring).

3.5. Synthesis of N^1,N^3,N^5-Tris(7-(1,2-dihydro-1,2-methanofullerene[60]-61-carbonyl)-9,9-di(methoxyethyl) fluoren-2-yl)-1",3",5"-tris(phenylamino)benzene) as Tris[(DPAF-C$_{2M}$)-C$_{60}$(>DPAF-C$_9$)] (4-C$_{2M-9}$)

The reagent 1,8-diazabicyclo[5.4.0]undec-7-ene (DBU, 0.21 g, 1.38 mmol) was added slowly to a homogeneous mixture of C$_{60}$>(DPAF-C$_9$) (1.1 g, 0.97 mmol) and cis-cup-tris(BrDPAF-C$_{2M}$) (cis-cup-3-C$_{2M}$, 0.31 g, 0.20 mmol) in anhydrous toluene (700 mL). During the Bingel reaction for the first 2.0 h, the products of mono- and bis-adducts became visible by the TLC technique, showing two brown bands at R_f = ~0.2 and ~0.3, respectively, using a mixture of toluene–EtOAc (7:3, v/v) as eluent. After a longer reaction period of 4 h, three brown bands close to each other were observed at R_f = ~0.2, ~0.3, and ~0.4 with the former band progressively becoming faint. At the end of the reaction (8.0 h), only two latter bands remained at R_f = ~0.3 and ~0.4 with the latter as the major chromatographic fraction. At this stage, the reaction mixture was concentrated to a 10% volume and then precipitated from methanol (100 mL) to afford the crude product mixture, which was isolated by centrifugation. Further purification by column chromatography (silica gel) using toluene to a solvent mixture of toluene–EtOAc (7:3, v/v) as the eluent with sequential increments of increasing solvent polarity afforded cis-cup-tris[(DPAF-C$_{2M}$)-C$_{60}$(>DPAF-C$_9$)]. It was further purified by TLC with isolation of only the narrow dense color fraction band to give brown solids of cis-cup-4-C$_{2M-9}$ in 79% yield (0.74 g) (at R_f = ~0.4 on TLC). Spectroscopic data: FT-IR (KBr) v_{max} 3062 (w, aromatic C-H stretching), 3031 (w), 3016 (w), 2952 (w, aliphatic C-H stretching), 2925 (m), 2852 (w), 1679 (s, C=O), 1591 (s, C=C), 1490 (s, anti-symmetric deformations of CH$_3$ groups and scissor vibrations of CH$_2$ groups), 1465 (s), 1419 (m), 1346 (w), 1315 (m), 1274 (s), 1238 (m), 1211 (s), 1170 (m), 1110 (s, stretching vibrations of C-O-C), 1072 (w), 1029 (w), 950 (w), 879 (w), 815 (m), 746 (s), 694 (s), 574 (w) and 524 (s, <C$_{60}$>) cm^{-1}; UV-vis (CHCl$_3$, 1.0 × 10^{-5} M) λ_{max} (ε) 296 nm (1.82 × 10^5 L mol^{-1} cm^{-1}) and 411 nm (1.11 × 10^5 L mol^{-1} cm^{-1}); ^{1}H NMR (500 MHz, CDCl$_3$) δ 8.65–8.12 (m, 12H, fluorenyl protons next to the keto group), 7.91–7.48 (m, 12H, fluorenyl protons), 7.31–7.06 [m, 57H, 45 aminophenyl protons and 12 fluorenyl protons (2H for each fluorene ring) next to N-atom), 6.57 (m, 3H, aromatic protons of the central phenyl ring), 5.78–5.25 (m, 6H, α-proton next on C$_{61}$), 3.01–2.81 (centered at δ 2.93) (m, 18H, primary C$_{2M}$ alkyl protons next to O-atom), 2.81–2.50 (centered at δ 2.65, m, 12H, secondary C$_{2M}$ alkyl protons next to O-atom), 2.48–2.07 (centered at δ 2.17, m, 12H, C$_{2M}$ alkyl protons next to the fluorine ring), 2.07–1.18 (m, 12H, C$_9$ alkyl protons next to the fluorene ring), 1.18–0.94 (centered at δ 1.14) (m, 6H, tertiary C$_9$ alkyl protons), 0.94–0.30 (centered at δ 0.70) (m, 96H, primary and secondary C$_9$ alkyl protons).

3.6. ROS Measurements Using singlet oxygen ($^1O_2$)-Sensitive Fluorescent Probe

The compound α,α'-(anthracene-9,10-diyl)bis(methylmalonic acid) (ABMA) was used as a fluorescent probe for singlet oxygen ($^1O_2$) trapping. The quantity of $^1O_2$ generated was monitored and counted by the relative intensity decrease of fluorescence emission of ABMA at 428 nm under excitation wavelengths of 380 nm (λ_{ex}). A typical probe solution was prepared by diluting a master solution of ABMA (1.0 × 10^{-5} M in DMF, 0.4 mL) with an amount 9-fold in volume of CHCl$_3$ (3.2 mL) in a cuvette (10 × 10 × 45 mm). The solution was added by a pre-defined volume of tris[(DPAF-C$_{2M}$)-C$_{60}$(>DPAF-C$_9$)] in CHCl$_3$ (1.0 × 10^{-5} M, 0.4 mL), followed by periodic illumination using an ultrahigh power white light LED lamp (Prizmatix, operated at the emission peak maxima centered at 451 and 530 nm with the collimated optical power output of >2.0 W in a diameter of 5.2 cm). Progressive fluorescent spectra were taken on the PTI QuantaMaster™ 40 Fluorescence Spectrofluorometer.

3.7. ROS Measurements Using Superoxide Radical ($O_2^-\cdot$)-Sensitive Fluorescent Probe

A superoxide radical ($O_2^-\cdot$)-reactive fluorescent probe, non-fluorescent potassium bis(2,4-dinitrobenzenesulfonyl)-2′,4′,5′,7′-tetrafluorofluorescein-10′ (or 11′)-carboxylate regioisomers (DNBs-TFFC), was used for the experiment. A typical probe solution [10^{-6} M in toluene–DMSO (9:1)] was prepared by diluting a stock solution of DNBs-TFFC in DMSO (10^{-5} M, 1.0 mL) by 10 times with toluene (9.0 mL). A dialysis film (CE) with the molecular weight cut-off (MWCO) of 100–500 Da was used to separate the solution of tris[(DPAF-C$_{2M}$)-C$_{60}$(>DPAF-C$_9$)] [10^{-6} M in toluene–DMSO (9:1)] from the probe solution kept in a cuvette with stirring during the fluorescent measurement. Only the solution of 4-C$_{2M}$ in the membrane sack was subjected to the LED light exposure at the excitation wavelength of 400–700 nm (white light). The quantity of $O_2^-\cdot$ generated was counted in association with its reaction with DNBs-TFFC that resulted in the product of highly fluorescent potassium 2′,4′,5′,7′-tetrafluorofluorescein-10′ (or 11′)-carboxylate regioisomers (TFFC) with fluorescence emission at 530 nm (λ_{em}) upon excitation at 484 nm (λ_{ex}). The detected intensity increase of fluorescence emission was then correlated to the relative quantity of O_2^- produced.

4. Conclusions

Previous studies on detected photoswitchable dielectric amplification phenomena by the simulated ferroelectric-like capacitor design using the construction and the fabrication of a fullerosome shell layer on core-shell γ-FeO$_x$@AuNPs were based on the photoinduced intramolecular charge-polarization of (C$_{60}$> acceptor)-(DPAF-C$_n$ donor) conjugates [16–18]. The corresponding formation of dielectric ion-radical components (C$_{60}$>)$^-$ and DPAF$^{+\cdot}$-C$_n$ within a fullerosome array layer was the basis of observed dielectric properties. Accordingly, several such conjugates were developed by the extension from the initial C$_{60}$(>DPAF-C$_9$) to demonstrate the correlation of the structural relationship and the chemical modifications to the enhanced dielectric properties, as stated above. Two interesting modifications both involved 3D-conformeric C$_{60}$(>DPAF-C$_9$) derivatives by fusing three phenyl rings of three diphenylamino groups together to form a shared central benzene moiety as the base of 3D configuration design. Specifically, the successful synthesis of *cis-cup*-tris[(DPAF-C$_{2M}$)-C$_{60}$(>DPAF-C$_9$)] stereoisomer may be beneficial for use as positive (DPAF-C$_n$)$^+$ and negative charge (<C$_{60}$>)$^-$ carriers in enhancing photoinduced dielectric characteristics [43]. It can also be applied as the precursor building block in the synthesis of several C$_{60}$- and C$_{70}$-based ultrafast photoresponsive nonlinear two-photon absorptive nanomaterials. Accordingly, we demonstrated efficient intramolecular energy and electron transfer capabilities of 3D conformer *cis-cup*-4-C$_{2M-9}$ using photophysical measurements and its effective production of singlet oxygen (via the energy transfer mechanism) and superoxide radicals (via the electron transfer mechanism). They can be applied as nano-photosensitizers [39–42] and nonlinear photonic agents [30–32,44].

Supplementary Materials: Supplementary Materials can be accessed online.

Author Contributions: All authors contributed significant effort on this work. H.Y. and M.W. carried out the main synthetic works, spectroscopic characterization, data analysis, and physical measurements; L.-S.T., and L.Y.C. participated in the discussion and experimental studies and contributed to a part of manuscript writing; All authors read and approved the final manuscript.

Funding: This research was funded by Air Force Office of Scientific Research (AFOSR) under the grant number FA9550-14-1-0153.

Conflicts of Interest: The authors declare no conflicts of interest.

References

1. El-Khouly, M.E.; Ito, O. Intermolecular and supramolecular photoinduced electron transfer processes of fullerene–porphyrin/phthalocyanine systems. *J. Photochem. Photobiol. C Photochem. Rev.* **2004**, *5*, 79–104.
2. Escudero, D. Revising intramolecular photoinduced electron transfer (PET) from first-principles. *Acc. Chem. Res.* **2016**, *49*, 1816–1824. [CrossRef] [PubMed]
3. Ito, O.; D'Souza, F. Recent advances in photoinduced electron transfer processes of fullerene-based molecular assemblies and nanocomposites. *Molecules* **2012**, *17*, 5816–5835. [PubMed]
4. Wu, W.; Zhao, J.; Sun, J.; Guo, S. Light-harvesting fullerene dyads as organic triplet photosensitizers for triplet–triplet annihilation upconversions. *J. Org. Chem.* **2012**, *77*, 5305–5312. [CrossRef] [PubMed]
5. Ziessel, R.; Allen, B.D.; Rewinska, D.B.; Harriman, A. Selective triplet-state formation during charge recombination in a fullerene/bodipy molecular dyad (bodipy=borondipyrromethene). *Chem. Eur. J.* **2009**, *15*, 7382–7393. [CrossRef] [PubMed]
6. Zhao, J.; Wu, W.; Sun, J.; Guo, S. Triplet photosensitizers: From molecular design to applications. *Chem. Soc. Rev.* **2013**, *42*, 5323–5351. [CrossRef]
7. Chea, Y.; Yuan, X.; Cai, F.; Zhaoa, J.; Zhaoa, X.; Xub, H.; Liu, L. Bodipy–corrole dyad with truxene bridge: Photophysical properties and application in triplet–triplet annihilation upconversion. *Dyes Pigments* **2019**, *171*, 107756. [CrossRef]
8. Kamkaew, A.; Lim, S.H.; Lee, H.B.; Kiew, L.V.; Chung, L.Y.; Burgess, K. BODIPY dyes in photodynamic therapy. *Chem. Soc. Rev.* **2013**, *42*, 77–88. [CrossRef]
9. Natali, M.; Campagna, S.; Scandola, F. Photoinduced electron transfer across molecular bridges: Electron- and hole-transfer superexchange pathways. *Chem. Soc. Rev.* **2014**, *43*, 4005–4018. [CrossRef]

10. Imahori, H.; Sakata, Y. Donor-linked fullerenes: Photoinduced electron transfer and its potential application. *Adv. Mater.* **1997**, *9*, 537–546.
11. D'Souza, F.; Ito, O. Photosensitized electron transfer processes of nanocarbons applicable to solar cells. *Chem. Soc. Rev.* **2012**, *41*, 86–96. [CrossRef] [PubMed]
12. Bottari, G.; Torre, G.; Guldi, D.M.; Torres, T. Covalent and noncovalent phthalocyanine-carbon nanostructure systems: Synthesis, photoinduced electron transfer, and application to molecular photovoltaics. *Chem. Rev.* **2010**, *110*, 6768–6816. [CrossRef] [PubMed]
13. Daly, B.; Ling, J.; Prasanna de Silva, A. Current developments in fluorescent PET (photoinduced electron transfer) sensors and switches. *Chem. Soc. Rev.* **2015**, *44*, 4203–4211. [CrossRef] [PubMed]
14. Yin, R.; Wang, M.; Huang, Y.-Y.; Chiang, L.Y.; Hamblin, M.R. Photodynamic therapy with decacationic [60]fullerene monoadducts: Effect of a light absorbing e^--donor antenna and micellar formulation. *Nanomed. Nanotechnol. Biol. Med.* **2014**, *10*, 795–808. [CrossRef] [PubMed]
15. Yin, R.; Wang, M.; Huang, Y.-Y.; Landi, G.; Vecchio, D.; Chiang, L.Y.; Hamblin, M.R. Antimicrobial photodynamic inactivation with decacationic functionalized fullerenes: Oxygen independent photokilling in presence of azide and new mechanistic insights. *Free Radic. Biol. Med.* **2014**, *79*, 14–27. [CrossRef] [PubMed]
16. Segura, J.L.; Martin, N. [60]Fullerene dimer. *Chem. Soc. Rev.* **2000**, *29*, 13–25. [CrossRef]
17. Shirai, Y.; Osgood, A.J.; Zhao, Y.; Kelly, K.F.; Tour, J.M. Directional control in thermally driven single-molecule nanocars. *Nano Lett.* **2005**, *5*, 2330–2334. [CrossRef]
18. Akimov, A.V.; Nemukhin, A.V.; Moskovsky, A.A.; Kolomeisky, A.B.; Tour, J.M. Molecular dynamics of surface-moving thermally driven nanocars. *J. Chem. Theory Comput.* **2008**, *4*, 652–656. [CrossRef]
19. Sasaki, T.; Osgood, A.J.; Kiappes, J.L.; Kelly, K.F.; Tour, J.M. Synthesis of a porphyrin-fullerene pinwheel. *Org. Lett.* **2008**, *10*, 1377–1380. [CrossRef]
20. Zhang, J.; Porfyrakis, K.; Morton, J.J.L.; Sambrook, M.R.; Harmer, J.; Xiao, L.; Ardavan, A.; Briggs, G.A.D.; Briggs, G. Photoisomerization of a fullerene dimer. *J. Phys. Chem.* **2008**, *C 112*, 2802–2904. [CrossRef]
21. Wang, J.L.; Duan, X.F.; Jiang, B.; Gan, L.B.; Pei, J.; He, C.; Li, Y.F. Nanosized rigid π-conjugated molecular heterojunctions with multi[6 0]fullerenes: Facile synthesis and photophysical properties. *J. Org. Chem.* **2006**, *71*, 4400–4410. [CrossRef] [PubMed]
22. López-Andarias, J.; Bauza, A.; Sakai, N.; Frontera, A.; Matile, S. Remote control of anion-π catalysis on fullerene-centered catalytic triads. *Angew. Chem.* **2018**, *130*, 11049–11053. [CrossRef]
23. Sabirov, D.S. Polarizability of C_{60} fullerene dimer and oligomers: The unexpected enhancement and its use for rational design of fullerene-based nanostructures with adjustable properties. *RSC Adv.* **2013**, *3*, 19430. [CrossRef]
24. Pankratyev, E.Y.; Tukhbatullina, A.A.; Sabirov, D.S. Dipole polarizability, structure, and stability of [2+2]-linked fullerene nanostructures $(C_{60})n$ ($n \leq 7$). *Phys. E Low-Dimens. Syst. Nanostruct.* **2017**, *86*, 237–242. [CrossRef]
25. Tukhbatullina, A.; Shepelevich, I.; Sabirov, D.S. Exaltation of polarizability as a common property of fullerene dimers with divers intercage bridges. *Fuller. Nanotub. Carbon Nanostruct.* **2018**, *26*, 661–666. [CrossRef]
26. Swart, M.; van Duijnen, P.T. Rapid determination of polarizability exaltation in fullerene-based nanostructures. *J. Mater. Chem.* **2015**, *C3*, 23–25. [CrossRef]
27. Wang, M.; Su, C.; Yu, T.; Tan, L.-S.; Hu, B.; Urbas, A.; Chiang, L.Y. Novel photoswitchable dielectric properties on nanomaterials of electronic core-shell γ-FeO_x@Au@fullerosomes for GHz frequency applications. *Nanoscale* **2016**, *8*, 6589–6599. [CrossRef]
28. Wang, M.; Yu, T.; Tan, L.-S.; Urbas, A.; Chiang, L.Y. Tunability of rf-responses by plasmonic dielectric amplification using branched e^--polarizable C_{60}-adducts on magnetic nanoparticles. *J. Phys. Chem. C* **2016**, *120*, 17711–17721. [CrossRef]
29. Wang, M.; Yu, T.; Tan, L.-S.; Urbas, A.; Chiang, L.Y. Enhancement of photoswitchable dielectric property by conducting electron donors on plasmonic core-shell gold-fluorenyl C_{60} nanoparticles. *J. Phys. Chem. C* **2018**, *122*, 12512–12523. [CrossRef]
30. Padmawar, P.A.; Rogers, J.O.; He, G.S.; Chiang, L.Y.; Canteenwala, T.; Tan, L.-S. Large cross-section enhancement and intramolecular energy transfer upon multiphoton absorption of hindered diphenylaminofluorene-C_{60} dyads and triads. *Chem. Mater.* **2006**, *18*, 4065–4074. [CrossRef]
31. Padmawar, P.A.; Canteenwala, T.; Tan, L.-S.; Chiang, L.Y. Synthesis and characterization of photoresponsive diphenylaminofluorene chromophore adducts of [60]fullerene. *J. Mater. Chem.* **2006**, *16*, 1366–1378. [CrossRef]

32. Luo, H.; Fujitsuka, M.; Araki, Y.; Ito, O.; Padmawar, P.; Chiang, L.Y. Inter- and intramolecular photoinduced electron-transfer processes between C_{60} and diphenylaminofluorene in solutions. *J. Phys. Chem. B* **2003**, *107*, 9312–9318. [CrossRef]
33. Hu, R.; Lager, E.; Aguilar-Aguilar, A.; Liu, J.; Lam, J.W.Y.; Sung, H.H.Y.; Williams, I.D.; Zhong, Y.; Wong, K.S.; Pena-Cabrera, E.; et al. Twisted intramolecular charge transfer and aggregation-induced emission of BODIPY derivatives. *J. Phys. Chem. C* **2009**, *113*, 15845–15853. [CrossRef]
34. Kang, N.-G.; Kokubo, K.; Jeon, S.; Wang, M.; Lee, C.-L.; Canteenwala, T.; Tan, L.-S.; Chiang, L. Synthesis and photoluminescent properties of geometrically hindered *cis*-tris(diphenyl-aminofluorene) as precursors to light-emitting devices. *Molecules* **2015**, *20*, 4635–4654. [CrossRef] [PubMed]
35. Lee, Y.-T.; Wang, M.; Kokubo, K.; Kang, N.-G.; Wolf, L.; Tan, L.-S.; Chen, C.-T.; Chiang, L. New 3D-stereoconfigurated *cis*-tris(fluorenylphenylamino)-benzene with large steric hindrance to minimize π–π stacking in thin-film devices. *Dyes Pigments* **2018**, *149*, 377–386. [CrossRef]
36. Chiang, L.Y.; Padmawar, P.A.; Canteenwala, T.; Tan, L.-S.; He, G.S.; Kannan, R.; Vaia, R.; Lin, T.-C.; Zheng, Q.; Prasad, P.N. Synthesis of C_{60}-diphenylaminofluorene dyad with large 2PA cross-sections and efficient intramolecular two-photon energy transfer. *Chem. Commun.* **2002**, 1854–1855. [CrossRef]
37. Jeon, S.; Wang, M.; Ji, W.; Tan, L.-S.; Cooper, T.; Chiang, L.Y. Broadband two-photon absorption characteristics of highly photostable fluorenyl-dicyanoethylenylated [60]fullerene dyads. *Molecules* **2016**, *21*, 647. [CrossRef]
38. Maeda, H.; Yamamoto, K.; Nomura, Y.; Kohno, I.; Hafsi, L.; Ueda, N.; Yoshida, S.; Fukuda, M.; Fukuyasu, Y.; Yamauchi, Y.; et al. A design of fluorescent probes for superoxide based on a nonredox mechanism. *J. Am. Chem. Soc.* **2005**, *127*, 68–69. [CrossRef]
39. Wang, M.; Huang, L.; Sharma, S.K.; Jeon, S.; Thota, S.; Sperandio, F.F.; Nayka, S.; Chang, J.; Hamblin, M.R.; Chiang, L.Y. Synthesis and photodynamic effect of new highly photostable decacationically armed [60]- and [70]fullerene decaiodide monoadducts to target pathogenic bacteria and cancer cells. *J. Med. Chem.* **2012**, *55*, 4274–4285. [CrossRef]
40. Wang, M.; Maragani, S.; Huang, L.; Jeon, S.; Canteenwala, T.; Hamblin, M.R.; Chiang, L.Y. Synthesis of decacationic [60]fullerene decaiodides giving photoinduced production of superoxide radicals and effective PDT-mediation on antimicrobial photoinactvation. *Eur. J. Med. Chem.* **2013**, *63*, 170–184. [CrossRef]
41. Sperandio, F.F.; Sharma, S.K.; Wang, M.; Jeon, S.; Huang, Y.-Y.; Dai, T.; Nayka, S.; de Sousa, S.C.O.M.; Chiang, L.Y.; Hamblin, M.R. Photoinduced electron-transfer mechanisms for radical-enhanced photodynamic therapy mediated by water-soluble decacationic C_{70} and $C_{84}O_2$ fullerene derivatives. *Nanomed. Nanotech. Biol. Med.* **2013**, *9*, 570–579. [CrossRef] [PubMed]
42. Huang, L.; Wang, M.; Huang, Y.-Y.; El-Hussein, A.; Chiang, L.Y.; Hamblin, M.R. Progressive cationic functionalization of chlorin derivatives for antimicrobial photodynamic inactivation and related vancomycin conjugate. *Photochem. Photobiol. Sci.* **2018**, *17*, 638–651. [CrossRef] [PubMed]
43. Yin, H.; Wang, M.; Yu, T.; Tan, L.-S.; Chiang, L.Y. 3D-conformer of tris[60]fullerenylated *cis*-tris(diphenylaminofluorene) as photoswitchable charge-polarizer on GHz-responsive trilayered core-shell dielectric nanoparticles. *Molecules* **2018**, *23*, 1873. [CrossRef] [PubMed]
44. Jeon, S.; Haley, J.; Flikkema, J.; Nalla, V.; Wang, M.; Sfeir, M.; Tan, L.-S.; Cooper, T.; Ji, W.; Hamblin, M.R.; et al. Linear and nonlinear optical properties of light-harvesting hybrid [60]fullerene triads and tetraads with dual NIR two-photon absorption characteristics. *J. Phys. Chem. C* **2013**, *117*, 17186–17195. [CrossRef] [PubMed]

Sample Availability: Sample of tris[(DPAF-C_{2M})-C_{60}(>DPAF-C_9)] is available from the authors.

© 2019 by the authors. Licensee MDPI, Basel, Switzerland. This article is an open access article distributed under the terms and conditions of the Creative Commons Attribution (CC BY) license (http://creativecommons.org/licenses/by/4.0/).

Article

Graphene Oxide Tablets for Sample Preparation of Drugs in Biological Fluids: Determination of Omeprazole in Human Saliva for Liquid Chromatography Tandem Mass Spectrometry

Zeynab Zohdi [1,2,3], Mahdi Hashemi [3], Abdusalam Uheida [1], Mohammad Mahdi Moein [2] and Mohamed Abdel-Rehim [1,2,*]

1. Functional Materials Division, Department of Applied Physics, School of Engineering Sciences, KTH Royal Institute of Technology, Isafjordsgatan 22, Kista, SE-164 40 Stockholm, Sweden; zohdi_zeynab@yahoo.com (Z.Z.); salm@kth.se (A.U.)
2. Department of Clinical Neuroscience, Centre for Psychiatry Research, Karolinska Institutet, SE-171 76 Solna, Sweden; Mohammad.moein@ki.s
3. Department of Chemistry, University of Bu-Ali Sina, Hamadan 65174, Iran; Mahdi.Hashemi@hotmail.com
* Correspondence: mohamed.abdel.rehim@ki.se or Mohamed.astra@gmail.com

Received: 28 January 2019; Accepted: 23 March 2019; Published: 27 March 2019

Abstract: In this study, a novel sort of sample preparation sorbent was developed, by preparing thin layer graphene oxide tablets (GO-Tabs) utilizing a mixture of graphene oxide and polyethylene glycol on a polyethylene substrate. The GO-Tabs were used for extraction and concentration of omeprazole (OME) in human saliva samples. The determination of OME was carried out using liquid chromatography-tandem mass spectrometry (LC–MS/MS) under gradient LC conditions and in the positive ion mode (ESI+) with mass transitions of m/z 346.3→198.0 for OME and m/z 369.98→252.0 for the internal standard. Standard calibration for the saliva samples was in the range of 2.0–2000 nmol L^{-1}. Limits of detection and quantification were 0.05 and 2.0 nmol L^{-1}, respectively. Method validation showed good method accuracy and precision; the inter-day precision values ranged from 5.7 to 8.3 (%RSD), and the accuracy of determinations varied from −11.8% to 13.3% (% deviation from nominal values). The extraction recovery was 60%, and GO-Tabs could be re-used for more than ten extractions without deterioration in recovery. In this study, the determination of OME in real human saliva samples using GO-Tab extraction was validated.

Keywords: graphene oxide; omeprazole; liquid chromatography; tandem mass spectrometry; saliva; GO-Tabs

1. Introduction

Omeprazole (OME) is a well-known drug that is used as a proton pump inhibitor to reduce the amount of acid produced in the stomach. OME is used in the treatment of all acid-related diseases, and it was introduced into the market for the first time by AstraZeneca under the brand name Losec® [1]. The pure S-enantiomer of OME was subsequently commercialized under the name Nexium®. OME is classified as an effective and safe medicine [2]. Gas chromatography (GC) and liquid chromatography (LC) with mass spectrometric (MS) detection have been used for the determination of OME, and solid phase extraction (SPE) has often been utilized as a sample preparation technique for extracting OME from plasma samples [3,4].

Saliva provides a simple and non-invasive sample source compared to blood. Numerous publications report that saliva can be a good choice for the determination of drugs for diagnostic purposes. Several research groups have studied methods for determining (i.e., screening) different substances in oral fluid [5–7]. Until now OME was measured in saliva and plasma using HPLC [8] and liquid chromatography-tandem mass spectrometry (LC–MS/MS) [9] with simple extraction methods. However, developing a simple and novel sample preparation technique for measuring the OME in saliva samples is a demanding issue.

Sample preparation is the first and, oftentimes, the main step in bioanalytical methods. As a result, there is an increasing demand for biological sample preparation techniques that are simple, inexpensive and environmentally friendly, and that permit acceptable recovery and selectivity [10]. During the last two decades, several sample preparation techniques have been developed and directed towards automation and on-line coupling. Two of these techniques are microextraction by packed sorbent (MEPS) and molecularly imprinted polymer tablets (MIP-Tabs), both of which were developed and introduced by researchers in the Abdel-Rehim group. MEPS is a miniaturization form of SPE that reduces significantly the required amounts of sorbent, solvent and sample volume. The MEPS technique has been utilized to extract and enrich analytes from different matrices such as water, plasma, urine and blood [11–15]. MIP-Tabs were prepared using a thin film of MIP polymer on a polyethylene substrate as a support material, and have been applied to the extraction of methadone in plasma samples and amphetamine in urine samples with good recovery and precision [16,17].

Nano-materials are of interest in the field of sample preparation due to their unique properties—in this case, high surface area compared to bulk and microscale materials. High surface area provides high adsorption capacity and high pre-concentration factor [16]. Additionally, nano-materials can possess high chemical stability and can be easily functionalized to increase selectivity. In our recent work, reduced graphene oxide (GO) materials as a solid phase in the MEPS method, was successfully applied for extraction and measurement of local anesthesia in plasma and saliva samples [18]. GO as a potential candidate in biomedical application presents a large surface area and proper dispersibility in most solvents due to the formation of hydrogen bonds between polar functional groups of GO surface and water molecules [19]. Nowadays, there is still no promising available information about the in vitro and in vivo toxicity of GO, but it is well approved that the preparation of GO with high purity is a key factor from a safety aspect [20,21].

Here for the first time, a combination of graphene oxide and polyethylene glycol was used to prepare novel graphene oxide tablets (GO-Tabs) that were evaluated for the extraction of omeprazole in human saliva samples. GO-Tabs is a novel, straightforward and effective sample preparation trend which will be presented here for extraction of OME from saliva samples and measurement by LC–MS/MS.

2. Materials and methods

2.1. Chemicals and Reagents

OME (racemate) and (S)-Lansoprazole (internal standard, IS) were obtained from Sigma-Aldrich (Steinheim, Germany), and GO was obtained from Sigma-Aldrich (St. Louis, MO, USA). The polyethylene material (PE), with a pore size of 0.2 μm, was a commercial material and was obtained from Sigma-Aldrich and its surface was not chemically modified. This material is already used as filter for aqueous solutions. The tablet form was prepared by a homemade tool. HPLC grade acetonitrile and methanol were purchased from Merck (Darmstadt, Germany). Analytical grade formic acid and ammonium hydroxide were obtained from Merck (Darmstadt, Germany). A Milli-Q Plus water purification system from Millipore (Bedford, MA, USA) was used for water purification.

2.2. Instrumentation

The LC system used in this study included two Shimadzu pumps (LC-10ADvp, Kyoto, Japan) and an autosampler (CTC-Pal, Analytics AG, Zwingen, Switzerland) with 50 μL sample loop. The separation column was 100 mm × 2.1 mm i.d. packed with 3.5 μm Zorbax Bonus-RP particles

(Agilent, Palo Alto, CA, USA). The LC mobile phase A was 0.1% formic acid in acetonitrile/water (0.5:99.5 v/v) and mobile phase B contained 0.1% formic acid in acetonitrile/water (80:20 v/v). The mobile phase gradient was 30% to 90% phase B in 5 min, with a final hold at 90% B for 2 min before resetting at 30% phase B. The mobile phase flow rate was held constant at 0.6 mL min^{-1}.

The MS instrumentation consisted of a triple quadrupole MS analyzer (Quatro-micro, Waters, Manchester, UK) equipped with an electrospray ionization source (ESI+) operated in the positive ion mode. The MS source block and desolvation temperatures were set at 150 °C and 350 °C, respectively. Nitrogen was utilized as curtain gas (950 L h^{-1}), and argon was utilized as the collision gas (collision energy of 10 eV for OME and 20 eV for lansoprazole). OME analysis in saliva samples was performed by using multiple reaction monitoring (MRM) transitions of m/z 346.0 > 198.0 for OME and m/z 369.9 > 252.0 for the IS with a dwell time of 0.2 s/transition. Peak-area ratios (OME/IS) were used for all calculations. Data analysis was performed using MassLynx software (version 4.1) obtained from Waters (Manchester, UK).

2.3. Preparation and Characterization of GO-Tabs

Polyethylene substrates in the tablet form (9 mm diameter × 2 mm thickness) were washed with HCL (1 M) and then NaOH (1 M) in an ultrasonic bath for 10 min, and then they were washed with water and dried at room temperature. Graphene oxide (20 mg) was added to 10 mL of acetonitrile and ultrasonicated for 30 min, and then 25 mg of polyethylene glycol (PEG) was added to the graphene oxide suspension. PEG was used to improve the interfacial adhesion between the GO nano-particles and the polyethylene tablet surface on which they were coated, and the dispersion of GO in the resultant film. Blank polyethylene tablets (10 total) were immersed in the GO-PEG suspension and ultrasonicated for 1–4 h; it was found that 3 h was the optimum time. After ultrasonication, the tablets were removed and placed in a freeze dryer overnight. Figure 1 shows a photograph of some prepared GO-Tabs. The GO-Tabs were chemically and mechanically stable.

Figure 1. Photograph of prepared graphene tablets (GO-Tabs).

2.4. Preparation of Stock, Standard and Quality Control Solutions

Two stock solutions (100 µM each) were prepared in methanol [one for preparing standard samples and the other for preparing quality control (QC) samples]. The standard and QC samples were prepared in blank pooled human saliva samples (n = 6). The concentrations of standard compounds in the saliva were in the range of 2.0–2000 nM. The QC samples were prepared in saliva at three concentration levels: Low (QCL, 6 nM), medium (QCM, 900 nM), and high (QCH, 1600 nM).

2.5. Sample Preparation

Standard samples in saliva were freshly prepared for each validation assay, while QC samples in saliva were prepared and stored at −20 °C until needed. A 200 µL volume of each sample was mixed with a 100 µL of the IS (1000 nM in methanol), then diluted with water (1:4) and centrifuged for 3 min. Then a GO-Tab was immersed in each saliva sample and shaken for 10 min, after which it was removed and washed with 200 µL of water. Then the analyte and internal standard were desorbed (i.e., extracted) by soaking in 1.0 mL of methanol for 1.0 min. The eluates were evaporated to dryness and redissolved in 200 µL of LC mobile phase. A 30 µL volume of the final sample solution was injected into the LC–MS/MS for analysis.

3. Results and Discussion

In this study, GO-Tabs were prepared and investigated for the extraction of OME in human saliva samples. Factors affecting the extraction performance, including desorption solution, extraction time, desorption time, sample pH, sample concentration and adsorption capacity, were investigated to obtain the best extraction/recovery efficiency.

3.1. GO-Tab Morphology Analysis

As described above, thin layers of GO-PEG were absorbed into the pores and surface of a polyethylene film and frozen overnight. The resultant GO-Tabs were 9 mm in diameter and 2 mm in thickness (Figure 1). SEM images before (Figure 2A) and after (Figure 2B) GO-PEG addition clearly show pores (Figure 2A) that become covered with the GO-containing polyethylene film (Figure 2B).

(A)

Figure 2. *Cont.*

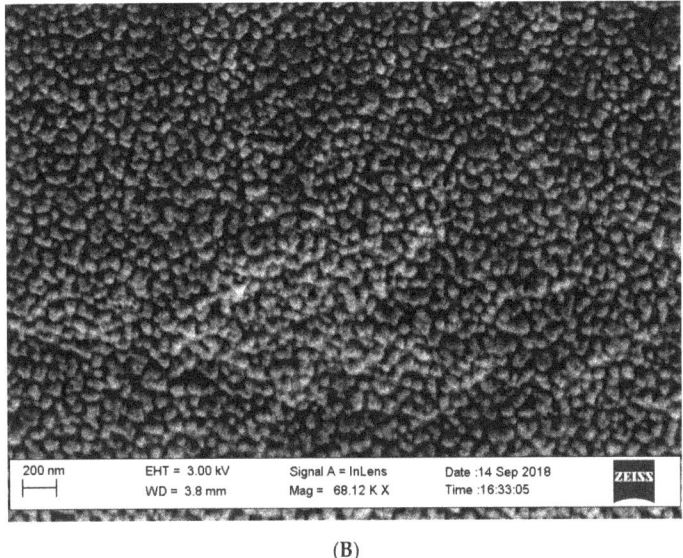

(B)

Figure 2. Electron micrographs (SEM) of GO-Tabs (**A**) before and (**B**) after polymerization.

3.2. Optimization of Extraction Protocol

3.2.1. Extraction Time

Because mass transfer is a time dependent process, the extraction time is an important factor. The effect of time on extraction recovery was investigated for 3.0, 5.0, 10.0, 20.0 and 30.0 min. The recovery was increased significantly when the time increased from 3.0 to 10.0 min (Figure 3A).

3.2.2. Desorption Time

The effect of desorption time on extraction efficiency was investigated for 1.0, 3.0 and 10.0 min, and the best result was obtained with 10.0 min (Figure 3B). After 10.0 min, no significant improvement was observed.

3.2.3. Type of Desorption Solvent

Desorption of extracted analyte from GO-Tabs was studied utilizing different solvents, including methanol, mixtures of methanol and water, and acetonitrile. Acetonitrile gave the highest recovery.

3.2.4. Effect of pH

The effect of sample pH on extraction recovery was investigated for three different pH values: Low (3), neutral (7) and high (12). The highest extraction yield was observed at neutral pH (Figure 3C).

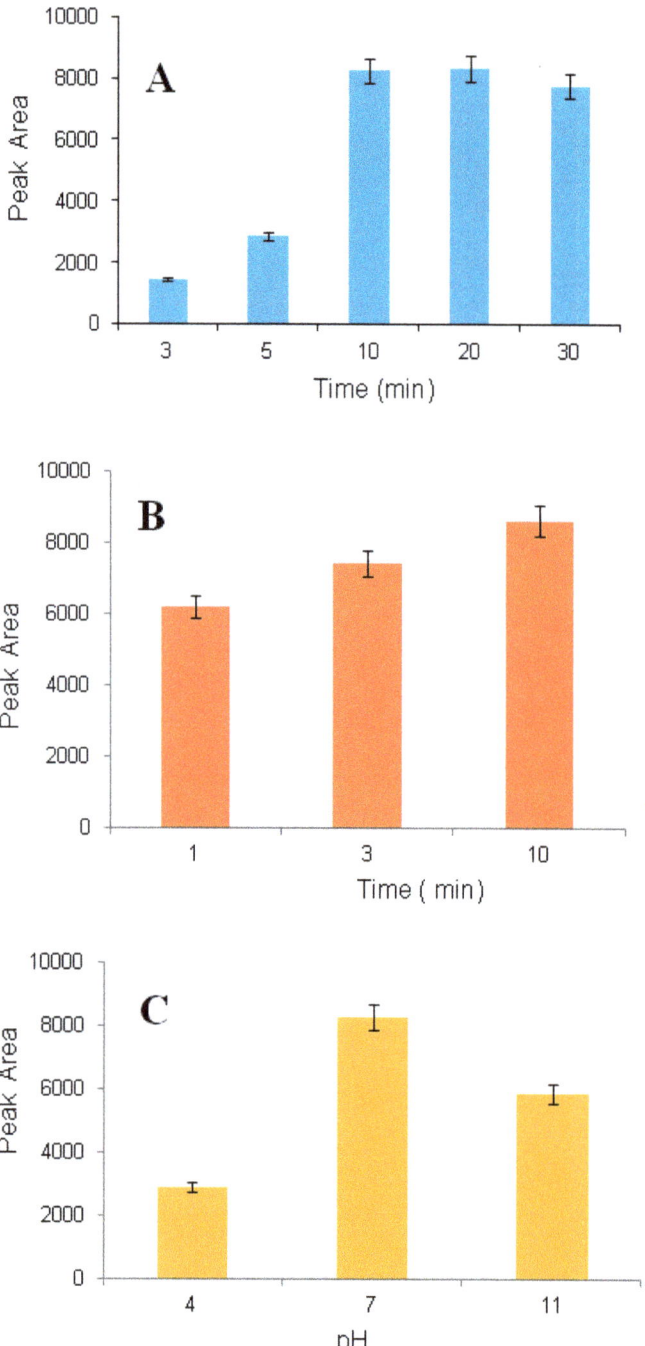

Figure 3. Effect of extraction time on extraction recovery (saliva sample, 1600 nmol L^{-1}) (**A**), effect of desorption time on extraction recovery (saliva sample, 1600 nmol L^{-1}) (**B**) and effect of sample pH (saliva sample, 1600 nmol L^{-1}) (**C**); pH was adjusted by addition of ammonium hydroxide and formic acid.

3.3. Adsorption Capacity

The adsorption capacity indicates the ability of the GO-Tabs to adsorb a specific analyte. In order to study the adsorption capacity of the GO-Tabs for OME, a series of different concentrations of OME in saliva samples ranging from 0.1 to 10 µmol L^{-1} were prepared. The extraction recovery was linear up to 6.0 µmol L^{-1}, and then the GO-Tab became saturated as shown in Figure 4.

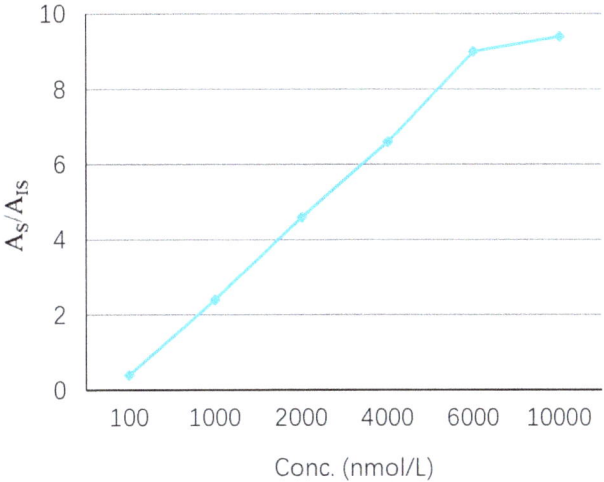

Figure 4. Capacity of GO-Tabs.

3.4. Selectivity of the GO-Tabs

The GO-Tab selectivity was investigated by comparing the extraction of OME in saliva using GO-Tabs and bare polyethylene tablets. The extraction recovery (recorded as area) using GO-Tabs was 4-fold higher compared with uncoated polyethylene tablets (Figure 5).

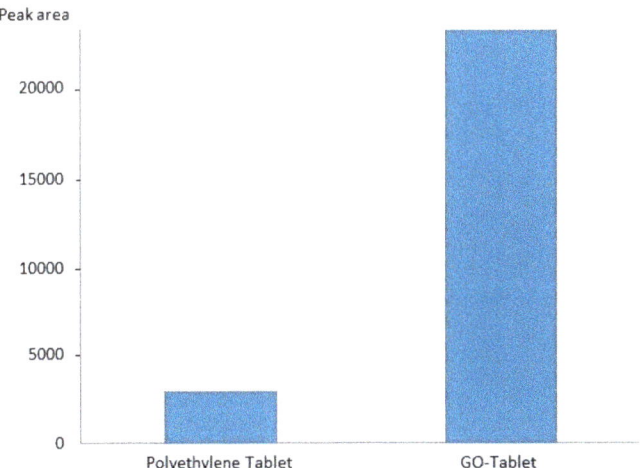

Figure 5. Efficiency of GO-Tabs and polyethylene blank tablet (saliva sample, 2000 nmol L^{-1}).

3.5. Method Validation

Validation of the method described in this study for determination of OME in human saliva was carried out according to international guidelines [22,23] and included linearity, limit of quantification (LLOQ), accuracy, precision, recovery, matrix effects, selectivity and carry-over.

3.5.1. Calibration, Selectivity and Extraction Efficiency

A standard calibration curve was constructed using eight OME standards prepared in saliva over the concentration range from 2.0 to 2000 nmol L^{-1}. A quadratic regression equation with $1/x$ weighting was used. The coefficient of determination (R^2) was 0.99 or higher for all analyses ($n = 3$). The limit of detection (LOD) was found to be 0.1 nmol L^{-1} and the limit of quantification (LLOQ) was equal to the lowest standard concentration (2.0 nmol L^{-1}) used to construct the calibration plot.

To evaluate the method selectivity, six different saliva samples were used. A saliva blank without OME and internal standard was analyzed and compared with a chromatogram obtained using a sample with OME concentration at the limit of quantification (LLOQ) to confirm the absence of endogenous interfering peaks at the same retention time of the OME analyte in the chromatograms. No significant peaks ($\geq$20% of the LLOQ) were observed at the same retention time as OME and the I.S. The method extraction recovery was found to be between 80 to 90%.

3.5.2. Accuracy and Precision

The relative error method (i.e., percent difference between the determined mean concentrations and the true concentrations) was used to evaluate accuracy, and the precision was calculated as the percentage of the relative standard deviation for the analysis of the quality control samples. For validation, three assays were done, and each assay consisted of eight calibration points and six quality control (QC) sample replicates at three concentration levels: Low (QCL: 6 nmol L^{-1}), medium (QCM: 900 nmol L^{-1}), and high (QCH: 1600 nmol L^{-1}). The accuracy was found to be in the range of 88.0–106.0% ($n = 18$), and intra- and inter-assay precisions were found to be in the range of 4.1–5.4% ($n = 6$) and 5.6–7.8% ($n = 18$), respectively (Table 1).

Table 1. Precision of quality control (QC) samples of omeprazole (OME) in human saliva.

Compound	Sample (Conc.)	Accuracy (%) ($n = 18$)	Precision (RSD%)	
			Intra-day ($n = 6$)	Inter-day ($n = 18$)
Omeprazole	QCL (6.0 nM)	105.5	5.4	7.8
	QCM (800 nM)	106.4	5.1	7.5
	QCH (1600 nM)	87.8	4.1	5.6

3.5.3. Method Selectivity and Matrix Effects

The method selectivity was examined by comparing LC–MS/MS chromatograms of the blank saliva sample (Figure 6A) and a standard sample spiked with internal standard and OME (Figure 6B). The blank saliva did not introduce any interfering peaks near the retention time of OME and the internal standard.

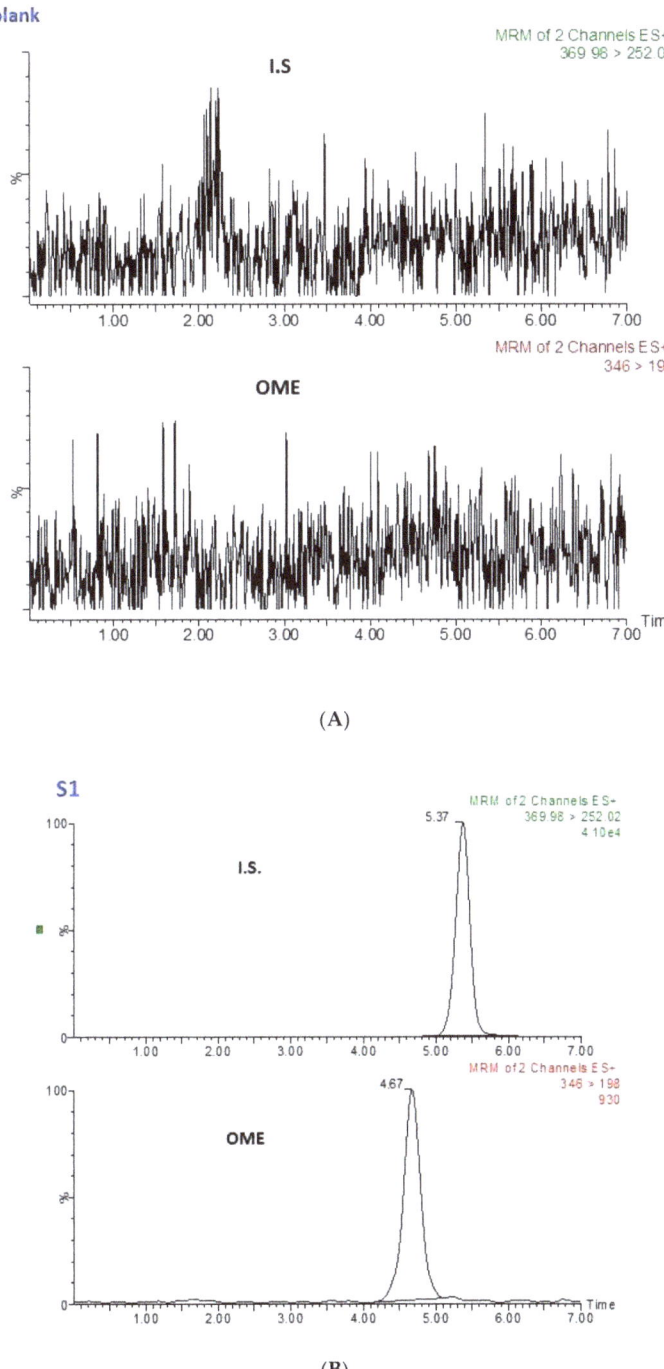

Figure 6. (**A**) Multiple reaction monitoring (MRM) transitions obtained from the analysis of a blank saliva sample, and (**B**) OME at 2 nmol L^{-1} (S1) with internal standard.

The effect of the saliva matrix on the MS signal was evaluated using the post-extraction addition method. Blank saliva was extracted according to the described protocol and OME was added to the extract at two concentration levels (QCL and QCH). Comparing LC–MS/MS analyses of these and pure methanol samples containing the same concentrations of OME showed that the saliva matrix did not affect the detector signal to any noticeable extent. Matrix effects ranged from −3% (using LQC) to −1% (using HQC).

3.5.4. Carry-Over and Reuse of GO-Tabs

After each extraction, the GO-Tab was washed first with methanol and then with water to eliminate carry-over into the next extraction. No carry-over could be detected when a blank saliva sample was extracted immediately after extraction of the highest concentration standard. A single GO-Tab could be re-used for ten extractions without any observable change in extraction efficiency ($n = 10$, SD = 2.4 and %RSD = 4.0).

4. Analysis of Patient Samples

The methodology developed in this study was used for the analysis of saliva samples from healthy subjects after administration of OME (20 mg dosage). Saliva samples were collected and analyzed for OME. Figure 7 shows the LC–MS/MS analysis of a patient saliva sample 2 h after administration of 20 mg of OME.

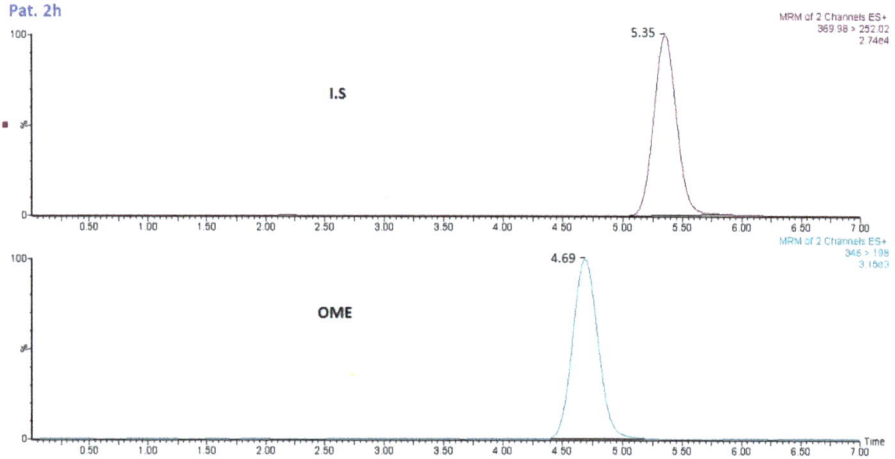

Figure 7. Chromatogram of OME from a patient sample (2 h after administration).

5. Conclusions

In this study, GO-Tabs were prepared using a novel sample preparation sorbent. GO was mixed with polyethylene glycol in acetonitrile and absorbed into a film of polyethylene using an ultrasonic bath. Polyethylene glycol was used to improve the dispersion of GO in the polyethylene substrate and the interfacial adhesion between GO and the substrate. This resulted in the formation of a layer of GO on the surface and within the pores of the polyethylene scaffold (Figures 1 and 2). Validation experiments demonstrated that this method accurately determined OME in human saliva samples with high precision and good sensitivity. The GO-Tabs could be re-used for at least up to ten extractions. GO-Tabs are a novel advance in sample preparation sorbent development and as a potential sorbent, it can be applied in other complex solutions in the near future.

Author Contributions: Z.Z.: labwork and validation; M.H.: co-supervision; A.U.: review and editing; M.M.M.: review and editing; M.A.-R.: principal investigator, main supervision, visualization.

Funding: This research received no external funding.

Conflicts of Interest: The authors declare no conflicts of interest.

References

1. Goodman, L.S.; Brunton, L.L.; Gilman, A.; Chabner, B.; Knollmann, B.C. *Goodman and Gilman's: The Pharmacological Basis of Therapeutics*; McGraw-Hill Medical: New York, NY, USA, 2011.
2. World Health Organization. *WHO Model Lists of Essential Medicines*; World Health Organization: Geneva, Switzerland, April 2015.
3. Cairns, A.M.; Chiou, R.H.Y.; Rogers, J.D.; Demetriades, J.L.J. Enantioselective high-performance liquid chromatographic determination of omeprazole in human plasma. *Chromatogr. B: Biomed. Sci. Appl.* **1995**, *666*, 323–328. [CrossRef]
4. Ishii, M.; Sato, M.; Ogawa, M.; Takubo, T.; Hara, K.i.; Ishii, Y. Simultaneous Determination of Omeprazole and its Metabolites (5′-Hydroxyomeprazole and Omeprazole Sulfone) in Human Plasma by Liquid Chromatography-Tandem Mass Spectrometry. *J. Liquid Chromatogr. Relat. Technol.* **2007**, *30*, 1797–1810. [CrossRef]
5. Pfaffe, T.; Cooper-White, J.; Beyerlein, P.; Kostner, K.; Punyadeera, C. Diagnostic potential of saliva: Current state and future applications. *Clinic. Chem.* **2011**, *57*, 675–687. [CrossRef] [PubMed]
6. Abdel-Rehim, A.; Abdel-Rehim, M. Screening and determination of drugs in human saliva utilizing microextraction by packed sorbent and liquid chromatography–tandem mass spectrometry. *Biomed. Chromatogr.* **2013**, *27*, 1188–1191. [CrossRef]
7. Liu, J.; Duan, Y. Saliva: A potential media for disease diagnostics and monitoring. *Oral Oncol.* **2012**, *48*, 569–577. [CrossRef]
8. Tamminga, W.J. Polymorphic Drug Metabolising Enzymes: Assessment of Activities by Phenotyping and Genotyping in Clinical Pharmacology. Ph.D. Thesis, University of Groningen, Groningen, The Netherlands, 2001.
9. Donzelli, M. Development, Validation and Application of the Basel Phenotyping Cocktail. Ph.D. Thesis, University of Basel, Basel, Switzerland, 2015.
10. Ashri, N.; Abdel-Rehim, M. Sample treatment based on extraction techniques in biological matrices. *Bioanalysis* **2011**, *3*, 1993–2008. [CrossRef]
11. Vita, M.; Abdel-Rehim, M.; Nilsson, C.; Hassan, Z.; Skansen, P.; Wan, H.; Meurling, L.; Hassan, M.J. Stability, pKa and plasma protein binding of roscovitine. *Chromatogr. B* **2005**, *821*, 75–80. [CrossRef]
12. Klimowska, A.; Wielgomas, B. Off-line microextraction by packed sorbent combined with on solid support derivatization and GC–MS: Application for the analysis of five pyrethroid metabolites in urine samples. *Talanta* **2018**, *176*, 165–171. [CrossRef]
13. Said, R.; Kamel, M.; El Beqqali, A.; Abdel-Rehim, M. Microextraction by packed sorbent for LC–MS/MS determination of drugs in whole blood samples. *Bioanalysis* **2011**, *29*, 197–205. [CrossRef]
14. Noche, G.G.; Laespada, M.E.F.; Pavón, J.L.P.; Cordero, B.M.; Lorenzo, S.M.; Noche, G.G.; Laespada, M.E.F.; Pavón, J.L.P.; Cordero, B.M.; Lorenzo, S.M. Determination of chlorobenzenes in water samples based on fully automated microextraction by packed sorbent coupled with programmed temperature vaporization–gas chromatography–mass spectrometry. *Anal. Bioanal. Chem.* **2013**, *405*, 6739–6748. [CrossRef] [PubMed]
15. Daryanavard, S.M.; Jeppsson-Dadoun, A.; Andersson, L.; Hashemi, M.; Colmsjö, A.; Abdel-Rehim, M. Molecularly imprinted polymer in microextraction by packed sorbent for the simultaneous determination of local anesthetics: lidocaine, ropivacaine, mepivacaine and bupivacaine in plasma and urine samples. *Biomed. Chromatogr.* **2013**, *27*, 1481–1488. [CrossRef] [PubMed]
16. El-Beqqali, A.; Abdel-Rehim, M. Molecularly imprinted polymer-sol-gel tablet toward micro-solid phase extraction: I. Determination of methadone in human plasma utilizing liquid chromatography–tandem mass spectrometry. *Anal. Chim. Acta* **2016**, *936*, 116–122. [CrossRef] [PubMed]

17. El-Beqqali, A.; Andersson, L.; Dadoun Jeppsson, A.; Abdel-Rehim, M.J. Molecularly imprinted polymer-sol-gel tablet toward micro-solid phase extraction: II. Determination of amphetamine in human urine samples by liquid chromatography–tandem mass spectrometry. *Chromatogr. B* **2017**, *1063*, 130–135. [CrossRef] [PubMed]
18. Ahmadi, M.; Moein, M.M.; Madrakian, T.; Afkhami, A.; Bahar, S.; Abdel-Rehim, M.J. Reduced graphene oxide as an efficient sorbent in microextraction by packed sorbent: Determination of local anesthetics in human plasma and saliva samples utilizing liquid chromatography-tandem mass spectrometry. *Chromatogr. B* **2018**, *1095*, 177–182. [CrossRef] [PubMed]
19. Ahmadi, M.; Madrakian, T.; Abdel Rehim, M. Nanomaterials as sorbents for sample preparation in bioanalysis: A review. *Anal. Chim. Acta.* **2017**, *958*, 1–21. [CrossRef] [PubMed]
20. Seabra, A.B.; Paula, A.J.; Lima, S.S.; Alves, O.L.; Durán, N. Nanotoxicity of graphene and graphene oxide. *Chem. Res. Toxicol.* **2014**, *27*, 159–168. [CrossRef] [PubMed]
21. Faria, A.F.; Martinez, D.S.T.; Moraes, A.C.M.; Maia da Costa, M.E.H.; Barros, E.B.; Souza Filho, A.G.; Paula, A.J.; Alvez, O.L. Unveiling the role of oxidation debris on the surface chemistry of graphene through the anchoring of Ag nanoparticles. *Chem. Mater.* **2012**, *24*, 4080–4087. [CrossRef]
22. *Guidance for Industry: Bioanalytical Method Validation*; U.S. Department of Health and Human Services, Food and Drug Administration, Center for Drug Evaluation and Research, Center for Veterinary Medicine: Rockville, MD, USA, 2001.
23. Moein, M.; El Beqqali, A.; Abdel-Rehim, M. Bioanalytical method development and validation: Critical concepts and strategies. *J. Chromatogr. B* **2017**, *1043*, 3–11. [CrossRef] [PubMed]

Sample Availability: Samples of the compounds are not available from the authors.

© 2019 by the authors. Licensee MDPI, Basel, Switzerland. This article is an open access article distributed under the terms and conditions of the Creative Commons Attribution (CC BY) license (http://creativecommons.org/licenses/by/4.0/).

MDPI
St. Alban-Anlage 66
4052 Basel
Switzerland
Tel. +41 61 683 77 34
Fax +41 61 302 89 18
www.mdpi.com

Molecules Editorial Office
E-mail: molecules@mdpi.com
www.mdpi.com/journal/molecules